NOBEL LECTURES IN PHYSICS

1991–1995

NOBEL LECTURES

INCLUDING PRESENTATION SPEECHES
AND LAUREATES' BIOGRAPHIES

PHYSICS

CHEMISTRY

PHYSIOLOGY OR MEDICINE

LITERATURE

PEACE

ECONOMIC SCIENCES

NOBEL LECTURES

INCLUDING PRESENTATION SPEECHES
AND LAUREATES' BIOGRAPHIES

PHYSICS

1991–1995

EDITOR

Gösta Ekspong

Department of Physics
University of Stockholm

Published for the Nobel Foundation in 1997 by

World Scientific Publishing Co. Pte. Ltd.
P O Box 128, Farrer Road, Singapore 912805
USA office: Suite 1B, 1060 Main Street, River Edge, NJ 07661
UK office: 57 Shelton Street, Covent Garden, London WC2H 9HE

NOBEL LECTURES IN PHYSICS (1991–1995)

ISBN 981-02-2677-2
ISBN 981-02-2678-0 (pbk)

Printed in Singapore by Uto-Print

FOREWORD

Since 1901 the Nobel Foundation has published annually "Les Prix Nobel" with reports from the Nobel award ceremonies in Stockholm and Oslo as well as the biographies and Nobel lectures of the laureates. In order to make the lectures available for people with special interests in the different prize fields the Foundation gave Elsevier Publishing Company the right to publish in English the lectures for 1901–1970, which were published in 1964–1972 through the following volumes:

Physics 1901–1970	4 vols.
Chemistry 1901–1970	4 vols.
Physiology or Medicine 1901–1970	4 vols.
Literature 1901–1967	1 vol.
Peace 1901–1970	3 vols.

Thereafter, and onwards the Nobel Foundation has given World Scientific Publishing Company the right to bring the series up to date and also publish the Prize lectures in Economics from the year 1969. The Nobel Foundation is very pleased that the intellectual and spiritual message to the world laid down in the laureates' lectures, thanks to the efforts of World Scientific, will reach new readers all over the world.

Bengt Samuelsson
Chairman of the Board

Michael Sohlman
Executive Director

Stockholm, October 1996

PREFACE

The lectures in physics by the Nobel laureates in 1991–1995 are collected in this volume. These lectures have been given at the Royal Swedish Academy of Sciences in Stockholm in the Nobel week of the year and originally published in a yearly volume *Les Prix Nobel*. The present volume also contains the presentation speeches held by a member of the Nobel Committee for physics during the prize-awarding ceremony on December 10 of each year. Included also are the short speeches by one of the physics laureates at the banquets on the evening of the same day.

Very interesting reading can be found in the autobiographies of each prizewinner. Some have started life without any thought of becoming a physicist and others have had a determined will to enter the research world of international physics. What influences are guiding a child or a youth turn out to be so diverse, that one can hardly write a recipe of how to develop into a physicist — let alone how to become a Nobel prizewinner.

The discoveries or inventions honoured by prizes during the five years covered here range from the smallest subatomic particles to one of the most extraordinary star systems in the sky.

In 1991 the prize to Pierre-Gilles de Gennes focused on his investigations of liquid crystals and polymers, materials of great practical importance and scientific interest.

The 1992 prize to Georges Charpak was given for his inventions and developments of charged particle detectors, in particular for his multiwire proportional counters, used in many high energy experiments the world over and more recently also applied to medicine.

In 1993 astrophysics came into focus by awarding the prize in physics to Russell Hulse and Joseph Taylor for their discovery of a close system of two stars, one of which is a fast rotating pulsar, which acts as an extraordinary accurate clock. The discovery story is a most fascinating one. The ultimate result turned out to be the best test of Einstein's theory of gravitation.

Pioneering contributions to neutron scattering techniques was the motivation for the physics prize in 1994 to Bertram Brockhouse and Clifford Shull. Neutrons, discovered in 1932, became available in large numbers after the war and were then tamed to serve science and technology. The methods worked out by the two prizewinners have been widely used and have become indispensible tools for the study and investigation of solid matter and liquids.

In 1995 the Academy focused on the smallest units of matter by honouring two achievements; the detection of the tiny neutrino was cited as motivation for the prize to Fredrick Reines and the discovery of the small but heavy tau-lepton for the prize to Martin Perl. Both particles are somewhat esoteric and remote from our every-day world, but both particles are believed to be of great importance for the life and development of the universe at large.

Gösta Ekspong

CONTENTS

Physics 1991

PIERRE-GILLES de GENNES

for discovering that methods developed for studying order phenomena in simple systems can be generalized to more complex forms of matter, in particular to liquid crystals and polymers

THE NOBEL PRIZE IN PHYSICS

Speech by Professor Ingvar Lindgren of the Royal Swedish Academy of Sciences.
Translation from the Swedish text.

Your Majesties, Your Royal Highnesses, Ladies and Gentlemen.

This year's Nobel Prize in Physics has been awarded to Pierre-Gilles de Gennes, Collège de France, Paris, for his investigations of *liquid crystals and polymers.* De Gennes has shown that mathematical models, developed for studying simpler systems, are applicable also to such complicated systems. De Gennes has discovered relations between different, seemingly quite unrelated, fields of physics — connections which noboby has seen before.

Liquid crystals and polymers can be regarded as intermediate states between *order and disorder.* A simple crystal, such as ordinary salt, is an example of almost perfect order — its atoms or ions are located in exact positions relative to each other. An ordinary liquid is an example of the opposite, complete disorder, its atoms or ions seem to move in completely irregular fashion. These examples represent two extremes of the concept order-disorder. In nature, there are more subtle forms of order and a liquid crystal is an example of that. It can be well ordered in one dimension but completely disordered in another. De Gennes has generalized the description of order for media of this type and been able to see analogies with, e.g. magnetic and superconducting materials.

The discovery of the remarkable substances we now call liquid crystals was made by the Austrian botanist *Friedrich Reinitzer* slightly more than a hundred years ago. In studying plants, he found that a substance related to the cholesterol had two distinct melting points. At the lower temperature, the substance became liquid but opaque and at the higher temperature completely transparent. Earlier, similar properties had been found in stearin. The German physicist *Otto Lehmann* found that the material was completely uniform between these temperatures with properties characteristic of a liquid as well as a crystal. Therefore, he named it "liquid crystal".

All of us have seen liquid crystals in the display of digital watches and pocket calculators. Most likely, we shall shortly see them also on the screen of our TV sets. Applications of this kind depend upon the unique optical properties of the liquid crystals and the fact that these can easily be changed, e.g. by an electric field.

It has been known for a long time that liquid crystals scatter light in an exceptional way, but all early explanations of this phenomenon failed. De Gennes found the explanation in the special way the molecules of a liquid crystal are ordered. One of the phases of a liquid crystal, called *nematic,* can be compared with a ferromagnet, where the atoms, which are themselves tiny magnets, are ordered so that they point in essentially the same direction – with slight variations. These variations follow a strict mathematical rule, which near the so-called *critical temperature,* where the magnet ceases to be magnetic, attains a very special form. In the liquid crystal the molecules are ordered in a similar way *at every temperature,* which explains its remarkable optical properties.

Another large field, where de Gennes has been very active, is that of *polymer physics.* A polymer consists of a large number of molecular fragments, monomers, which are linked together to form long chains or other structures. These molecules can be formed in a countless number of ways, giving the polymer materials a great variety of chemical and physical properties. We are quite familiar with some of the applications, which range from plastic bags to parts of automobiles and aircraft.

Also in these materials, de Gennes has found analogies with critical phenomena appearing in magnetic and superconducting materials. For instance, the size of the polymer in a solution increases by a certain power of the number of monomers, which is mathematically analogous to the behavior near a critical temperature of a magnet. This had led to the formulation of *scaling laws,* from which simple relations between different properties of polymers can be deduced. In this way, predictions can be made about unknown properties – predictions which later in many cases have been confirmed by experiments.

Major progress in science is often made by transfering knowledge from one discipline to another. Only few people have sufficiently deep insight and sufficient overview to carry out this process. De Gennes is definitely one of them.

Professor de Gennes,
You have been awarded the 1991 Nobel Prize in Physics for your outstanding contributions to the understanding of liquid crystals and polymers. It is my privilege to convey to you the heartiest congratulations of the Royal Swedish Academy of Sciences, and I now ask you to receive the Prize from the hands of His Majesty the King.

Pierre Gilles de Gennes

PIERRE-GILLES de GENNES

P. G. de Gennes was born in Paris, France, in 1932. He majored from the Ecole Normale in 1955. From 1955 to 1959, he was a research engineer at the Atomic Energy Center (Saclay), working mainly on neutron scattering and magnetism, with advice from A. Herpin, A. Abragam and J. Friedel (PhD 1957). During 1959 he was a postdoctoral visitor with C. Kittel at Berkeley, and then served for 27 months in the French Navy. In 1961, he became assistant professor in Orsay and soon started the Orsay group on *supraconductors*. Later, 1968, he switched to *liquid crystals*. In 1971, he became Professor at the Collège de France, and was a participant of STRASACOL (a joint action of Strasbourg, Saclay and Collège de France) on *polymer physics*.

From 1980, he became interested in interfacial problems, in particular the *dynamics of wetting*. Recently, he has been concerned with the physical chemistry of *adhesion*.

P.G. de Gennes has received the Holweck Prize from the joint French and British Physical Society; the Ampère Prize, French Academy of Science; the gold medal from the French CNRS; the Matteuci Medal, Italian Academy; the Harvey Prize, Israel; the Wolf Prize, Israel; The Lorentz Medal, Dutch Academy of Arts and Sciences; and polymer awards from both APS and ACS.

He is a member of the French Academy of Sciences, the Dutch Academy of Arts and Sciences, the Royal Society, the American Academy of Arts and Sciences, and the National Academy of Sciences, USA.

SOFT MATTER

Nobel Lecture, December 9, 1991

by

PIERRE-GILLES DE GENNES

Collège de France, Paris, France

What do we mean by soft matter? Americans prefer to call it "complex fluids". This is a rather ugly name, which tends to discourage the young students. But it does indeed bring in two of the major features:

1) Complexity. We may, in a certain primitive sense, say that modern biology has proceeded from studies on simple model systems (bacterias) to complex multicellular organisms (plants, invertebrates, vertebrates...). Similarly, from the explosion of atomic physics in the first half of this century, one of the outgrowths is soft matter, based on polymers, surfactants, liquid crystals, and also on colloidal grains.

2) Flexibility. I like to explain this through one early polymer experiment, which has been initiated by the Indians of the Amazon basin : they collected the sap from the hevea tree, put it on their foot, let it "dry" for a short time. And, behold, they have a *boot.* From a microscopic point of view, the starting point is a set of independent, flexible polymer chains. The oxygen from the air builds in a few bridges between the chains, and this brings in a spectacular change: we shift from a liquid to a network structure which can resist tension — what we now call a *rubber* (in French: caoutchouc, a direct transcription of the Indian word). What is striking in this experiment, is the fact that a very mild chemical action has induced a drastic change in mechanical properties : a typical feature of soft matter.

Of course, with some other polymer systems, we tend to build more rigid structures. An important example is an enzyme. This is a long sequence of aminoacids, which folds up into a compact globule. A few of these aminoacids play a critical role : they build up the "active site" which is built to perform a specific form of catalysis (or recognition). An interesting question, raised long ago by Jacques Monod, is the following : we have a choice of twenty aminoacids at each point in the sequence, and we want to build a receptor site where the active units are positioned in space in some strict way. We cannot just put in these active units, because, if linked directly, they would not realise the correct orientations and positions. So, in between two active units, we need a "spacer", a sequence of aminoacids which has enough variability to allow a good relative positioning of the active sites at both ends of the spacer. Monod's question was; what is the minimum length of spacers?

It turns out that the answer is rather sharply defined(1). The magic number is around 13—14. Below 14 units, you will not usually succeed in getting the desired conformation. Above 14, you will have many sequences which can make it. The argument is primitive; it takes into account excluded volume effects, but it does not recognise another need for a stable enzyme – namely that the interior should be built preferably with hydrophobic units, while the outer surface must be hydrophilic. My guess is that this cannot change the magic number by much more than one unit. Indeed, when we look at the spacer sizes in a simple globular protein like myosin, we see that they are not far from the magic number.

Let me return now to flexible polymers in solution, and sketch some of their strange mechanical properties. One beautiful example is the four roller experiment set up by Andrew Keller and his coworkers(2). Here, a dilute solution of coils is subjected to a purely longitudinal shear. If the exit trajectory is well chosen (in the symmetry plane of the exit channel), the molecules are stressed over long times. What is found is that, if the shear rate $\dot{\gamma}_c$ exceeds a certain threshold value $\mathring{\gamma}_c$, an abrupt transition takes place, and the medium becomes birefringent. This is what I had called a "coil-stretch transition"(3). When the shear begins to open the coil, it offers more grip to the flow, and opens even more... leading to a sharp transition. Here, we see another fascinating aspect of soft matter – the amazing coupling between mechanics and conformations. Indeed, Keller showed that rather soon (at shear rates $\dot{\gamma}_c > \mathring{\gamma}_c$, the chains break), and they do so very near to their midpoint – a spectacular result.

Another interesting feature of dilute coils is their ability to reduce the losses in turbulent flows. This is currently called the Toms effect. But in actual fact it was found, even before Toms, by Karol Mysels(4). He is here today, to my great pleasure. Together with M. Tabor, we tried to work out a scaling model of coils in a turbulent cascade(5), but our friends in mechanics think that it is not realistic, the future will tell what the correct answer is.

I have talked a lot about polymers. It would be logical to do the same with colloïds, or – as I like to call it – "ultra divided matter". But since I just gave another talk with this title at the Nobel symposium in Göteborg, I will omit the subject, in spite of its enormous practical importance.

Let me rather switch to *surfactants*, molecules with two parts: a polar head which likes water, and an aliphatic tail which hates water. Benjamin Franklin performed a beautiful experiment using surfactants; on a pond at Clapham Common, he poured a small amount of oleic acid, a natural surfactant which tends to form a dense film at the water-air interface. He measured the volume required to cover all the pond. Knowing the area, he then knew the height of the film, something like three nanometers in our current units. This was to my knowledge the first measurement of the size of molecules. In our days, when we are spoilt with exceedingly complex toys, such as nuclear reactors or synchrotron sources, I particularly like to describe experiments of this Franklin style to my students.

Surfactants allow us to protect a water surface, and to generate these beautiful soap bubbles, which are the delight of our children. Most of our understanding of these soap bubbles is due to a remarkable team, Mysels, Shinoda and Frankel, who wrote *the* book on this subject(6). Unfortunately, this book is now very hard to find, I very much hope that it will be reprinted.

Long ago Françoise Brochard, Jean-François Lennon and I(7) became interested in some bilayer systems, where we have two sheets of surfactant, each pointing towards the neighbouring water. A related (although more complex) system of this type is a red blood cell. For many years it had been known that, when observed under phase contrast, these cells *flicker*. – It was sometimes believed that this flicker reflected an instability of a living system under non-equilibrium conditions. Ultimately, the thing is simpler. The essential property of insoluble bilayers is that they optimise their area at fixed surfactant number. Thus, the energy is stationary with respect to area: the surface tension vanishes. This means that the fluctuations in shape of these deflated cells, or "vesicles", are huge: the flicker is just an example of Brownian motion for a very flexible object. What Jean-François had done was to measure space time correlations for the flicker. Françoise then showed that they could be understood from a model containing no surface tensions, but only curvature energies plus viscous forces – another good example of soft matter.

This was, in fact, one of the starting points for many studies on surfactant bilayers, pioneered by W. Helfrich and, on a more formal side, on random surfaces especially with D. Nelson. One of the great successes in this field has been the invention of the "sponge phase" of microemulsions(8,9). But, more generally, it is amusing to learn from these people that there is some overlap in thought between the highbrow string theories and the descriptions of soaps!

Let me now move to another corner in our garden – liquid crystals. Here, I must pay tribute first to two great pioneers:

i) Georges Friedel, who was the first to understand exactly what is a liquid crystal, and what are the main types; ii) Charles Frank, who (after some early work of Oseen) constructed the elastic theory of nematics, and described also a number of their topological defects ("disclinations").

I will talk here only about the smectics. Observing certain defects ("focal conics") in smectics, Friedel was able to prove that their structure must be a set of liquid, equidistant, deformable layers(10). By observations at the one hundred micron scale, he was thus able to infer the correct stucture at the ten Å scale – an amazing achievement.

Smectics bring me naturally to another important feature of complex fluids – namely that, in our days, it is sometimes possible *to create new forms of matter*. The sponge phase quoted above was an example. Another striking case was the invention of ferroelectric smectics by R.B. Meyer, in Orsay, circa 1975. He thought about a certain molecular arrangement, with chiral molecules, which should automatically generate a phase (the "C* phase") carrying a non-zero electric dipole. Within a few months, our local chemists had produced the right molecule, and the first liquid ferroelectric was born!(11). In our days, these materials may become very important for display purposes, they commute 10^3 times faster than the nematics in our wrist-watches.

Another case of far smaller importance, but amusing, is the "*ferrosmectic*" constructed by M. Veyssié and P. Fabre. The starting point is a water based ferrofluid; a suspension of very fine magnetic particles. (Ferrofluids were invented long ago by R. Rosensweig, and have many amazing properties). Here, what is done, is to prepare a "club sandwich" bilayer | ferrofluid | bilayer | ... A system like this, subjected to a magnetic field $\vec{H}$, is happier when $\vec{H}$ is parallel to the layers. It is then interesting to observe the sandwich, with a polarizing microscope, in the frustrated situation where $\vec{H}$ is normal to the layers. At very small H, nothing is seen. But beyond a certain weak threshold H_c, figures like flowers grow in the field(12). We understand this as a two step process a) just above threshold there is a chemical undulation instability b) later, focal conics appear, with a basic size imposed by the original undulation, but also with smaller conics (which are required to fill space correctly). This "club sandwich" is ultimately detecting rather weak magnetic fields ($\sim$ 30 gauss).

Let me quote still another new animal : the *Janus grains*, first made by C. Casagrande and M. Veyssié. The god Janus had two faces. The grains have two sides : one apolar, and the other polar. Thus, they have certain features in common with surfactants. But there is an interesting difference if we consider the films which they make, for instance at a water | air interface. A dense film of a conventional surfactant is quite impermeable. On the other hand, a dense film of Janus grains always has some interstices between the grains, and allows for chemical exchange between the two sides; "the skin can breathe". This may possibly be of some practical interest.

The first technique used to make the Janus grains was based on spherical particles, half embedded in a plastic and silanated on the accessible side(13). This produces only microquantities of material. But a group at Goldschmidt(14) research invented a much more clever pathway. The starting point is a collection of *hollow* glass particles, which are available commercially. There the outer surface is hydrophobized, and finally the particles are crushed. The resulting platelets have one side hydrophilic and one side hydrophobic. They are irregular, but they can be produced in tons.

I would like now to spend a few minutes thinking about the style of soft matter research. One first, major, feature, is the possibility of very simple experiments – in the spirit of Benjamin Franklin. Let me quote two examples. The first concerns the *wetting of fibers.* Usually a fiber, after being dipped in a liquid, shows a string of droplets, and thus, for some time, people thought that most common fibers were non-wettable. F. Brochard analysed theoretically the equilibria on curved surfaces, and suggested that in many cases we should have a wetting film on the fiber, in between the droplets. J.M. di Meglio and D. Queré established the existence, and the thickness, of the film, in a very elegant way(15). They created a pair of neighbouring droplets, one small and one large, and showed that the small one emptied slowly into the big one (as capillarity wants it to go). Measuring the speed of the process, they could go back to the thickness of the film which lies on the fiber and connects the two droplets : the Poiseuille flow rates in the film are very sensitive to thickness.

Another elegant experiment in wetting concerns the *collective modes of a contact line*; the edge of a drop standing on a solid. If one distorts the line by some external means, it returns to its equilibrium shape with a relaxation rate dependent upon the wavelength of the distortion, which we wanted to study. But how could we distort the line? I thought of very complex tricks, using electric fields from an evaporated metal comb, or other, even worse, procedures. But Thierry Ondarcuhu came up with a simple method.

1) He first prepared the unperturbed contact line L by putting a large droplet on a solid.

2) He then dipped a fiber in the same liquid, pulled it out, and obtained, from the Rayleigh instability, a very periodic string of drops.

3) He laid the fiber on the solid, parallel to L, and generated a line of droplets on the solid.

4) He pushed the line L (by tilting the solid), up to the moment where L touched the droplets; then coalescence took place, and he had a single, wavy line on which he could measure relaxation rates(16).

I have emphasized experiments more than theory. Of course we need some theory when thinking of soft matter. And in fact some amusing theoretical analogies sometimes show up between soft matter and other fields. One major example is due to S.F. Edwards(17). Edwards showed a beautiful correspondence between the conformations of a flexible chain and the trajectories of a non relativistic particle; the statistical weight of the chain corresponding to the propagator of the particle. In the presence of external potentials, both systems are ruled by exactly the same Schrödinger equation! This observation has been the key to all later developments in polymer statistics.

Another amusing analogy relates the smectics A to superconductors. It was discovered simultaneously by the late W. McMillan (a great scientist, who we all miss) and by us. Later, it has been exploited artistically by T. Lubensky and his colleagues(18). Here again, we see a new form of matter being invented. We knew that type II superconductors let in the magnetic field in the form of quantized vortices. The analog here is a smectic A inside which we add chiral solutes, which play the role of the field. In some favorable cases, as predicted in 1988 by Lubensky, this may generate a smectic phase drilled by screw dislocations — the so called A* phase. This was discovered experimentally only one year later by Pindak and coworkers(19), a beautiful feat.

Let me now end up this sentimental journey into soft matter, with a brief mention of my companions. Some were met during the way, like Jean Jacques, a great inventor of liquid crystals, or Karol Mysels, the undisputed master of surfactant science. Some others were with me all along the way; Henri Benoit and Sam Edwards, who taught me polymer science; Jacques des Cloizeaux and Gérard Jannink, who have produced a deep theoretical book on this subject. Finally, an inner core of fellow travelers, over all forms of land and sea : Phil Pincus, Shlomo Alexander, Etienne Guyon, Madeleine Veyssié; and last but not least, Françoise Brochard — sans laquelle les choses ne seraient que ce qu'elles sont.

The final lines are not mine: they come from an experiment on soft matter, after Boudin, which is shown on the following figure.

An English translation might run like this:

> "Have fun on sea and land
> Unhappy it is to become famous
> Riches, honors, false glitters of this world
> All is but soap bubbles"

No conclusion could be more appropriate today.

REFERENCES:

1. P G de Gennes, in Pvor. 2nd Conf. "Physique théorique et Biologie". Editions CNRS 1969 (15 Quai A. France 75007 Paris, France).
2. J.A. Odell, A. Keller, in *Polymer-flow Interactions* (ed. I. Rabin), AIP, New York 1985 ; Av. Keller, J. Odell, Coll. Polym. Sci., *263,* 181 (1985).
3. P.G. de Gennes, J. Chem. Phys., *60,* 5030 (1974).
4. For a historical review see K. Mysels, Chem. Eng. Prog., Symposium series, *67,* 45, (1971).
5. M. Tabor, P.G. de Gennes, Europhys. Lett., *2,* 519 (1986) P.G. de Gennes, Physica, *140 A,* 9 (1986).
6. K. Mysels, K. Shinoda, S. Frankel, *Soap Films*, Pergamon, London (1959).
7. F. Brochard, J.F. Lennon, J. Physique (Paris), *36,* 1035 (1976).
8. G. Porte, J. Marignan, P. Bassereau, R. May, J. Physique (Paris), *49,* 511 (1988).
9. D. Roux, M.E. Cates, Proceedings of the 4th Nishinomya—Yukawa Symposium, Springer (to be published).
10. G. Friedel, Annales de Physique, *18,* 273 (1922).
11. R.B. Meyer, L. Liebert, L. Strzelecki, P. Keller, J. Physique L. 69 (1975).
12. P. Fabre, C. Casagrande, M. Veyssié, V. Cabuil, R. Massart, "Ferrosmectics : A new Magnetic and Mesomorphic Phase", Phys. Rev. Lett., *64,* 539 (1990).
13. C. Casagrande, M. Veyssié, C.R. Acad. Sci. (Paris), *306 II,* 1423 (1988). C. Casagrande, P. Fabre, M. Veyssié, E. Raphaël, Europhys. Lett., *9,* 251 (1989).
14. B. Grüning, U. Holtschmidt, G. Koerner, G. Rössmy US Patent no 4, 715, 986 (Dec 1987).
15. J.M. di Meglio, C.R. Acad. Sci. (Paris), *303 II,* 437 (1986).
16. T. Ondarcuhu, M. Veyssié, Nature, *352,* 418 (1991).
17. S.F. Edwards, Proc. Phys. Soc. (London), *85,* 613 (1965).
18. S.R. Renn, T. Lubensky, Phys. Rev., *A 38,* 2132 (1988).
19. J.W. Goodby, M.A. Waugh, S.M. Stein, E. Chin, R. Pindak, J.S. Patel, J. Am. Chem. Soc., *111,* 8119 (1989).

Physics 1992

GEORGES CHARPAK

for his invention and development of particle detectors, in particular the multiwire proportional chamber

THE NOBEL PRIZE IN PHYSICS

Speech by Professor Carl Nordling of the Royal Swedish Academy of Sciences.
Translation from the Swedish text.

Your Majesties, Your Royal Highnesses, Ladies and Gentlemen,

This year the Nobel Prize in Physics has been awarded to Georges Charpak, France, for his invention and development of particle detectors, in particular the multiwire proportional chamber. It is the tenth time in the history of the Nobel Prize that the word "invention" has been used in the citation for the award in physics.

None of us owns the kind of detector for which the prize is being awarded today, but we are all equipped with other forms of detectors. Our eyes are detectors of light, our ears detect sound, our noses detect odors and so on. The signals from these sense organs are sent to a computer—the brain. There they are processed, communicated to our consciousness and used as the basis of our actions and our conception of the world in which we live.

But we are not always content with this. Our curiosity about the world extends beyond our immediate sensory impressions. For this reason inventive people have constructed devices of various kinds which intensify our senses or replace them completely—if this is at all possible in principle. Galileo Galilei constructed telescopes, Zacharias Janssen invented the microscope etc.

Today's elementary particle physicists look deep inside matter using accelerators as microscopes. In these accelerators particles chosen as suitable projectiles, electrons for instance, are raised to high energies and then made to collide with each other. This produces new particles like the sparks from fireworks. In this invisible deluge of sparks, which can be discharged a hundred million times each second, there is information about the innermost constituents of matter and the forces with which they interact.

In order to acquire this information, enormous installations are built, which contain various kinds of detectors. Professor Charpak has invented the detector which has meant most for the progress in the area of elementary particle physics during the last few decades.

The list of qualities demanded of a detector of elementary particles is a long one. It must react quickly, must be able to cover large surfaces—hundreds of square meters—and must send its signals direct to a computer. It must be sensitive to position, i.e. it must not only be able to say *if* something has happened but also *where,* and it must also be capable of following the total length of the trajectory of a particle, often several meters. And it must be able to do all this even when it is placed in a strong magnetic field.

All of these requirements are fulfilled by the multiwire proportional chamber, the detector which Georges Charpak invented in 1968. This detector is used, in some form or other, in more or less every experiment within elementary particle physics today, and Georges Charpak has been at the centre of the development which has taken place since the original invention was made. Many important discoveries have been made using his detectors.

Charpak's research is an example of an advanced technological development within basic science. Its original purpose was to contribute to the development of nuclear physics and elementary particle physics in order to provide further facets of our conception of the world. This aim has been achieved, but Charpak's detector has also found applications well outside the field of elementary particle physics, for instance in medicine. In this development too, Charpak plays a central role.

Monsieur Charpak,
Le Prix Nobel de Physique de l'année 1992 vous a été decerné pour votre invention et développement de détecteurs de particules, notamment de la chambre proportionelle multifils. J'ai l'honneur de vous adresser les félicitations les plus chaleureuses de l'Académie Royale des Sciences de Suède, et je vous invite à recevoir votre Prix des mains de Sa Majesté le Roi.

GEORGES CHARPAK

Né le ler août 1924 à Dabrovica (Pologne)
Naturalisé français en 1946

Etudes	
	Lycée Saint-Louis à Paris
	Lycé de Montpellier
1945–1947	Ecole des Mines de Paris
Diplômes	
1948	Licence des sciences, Ingénieur civil des mines
1954	Docteur en physique, Recherche expérimentale physique nucléaire au Collège de France
Carrière	
1948–1959	Au Centre National de la Recherche Scientifique (CNRS)
1959–1991	Au Centre Européen pour la Recherche Nucléaire (CERN)
Travaux	(Extraits)
1960	Participation à la première mesure précise du moment magnétique du muon
1962–1967	Invention de divers types de chambres à étincelles sans photographie (division de courant et retard des impulsions)
1962–1967	Etudes de structure nucléaire par les réactions (π^{+}2p)
1968	Introduction des chambres proportionnelles multifils et des chambres à dérive
1974	Introduction de la chambre à dérive sphérique pour l'étude des structures de proteines par diffraction des rayons X (ORSAY)
1979–1989	Introduction des chambres à avalanches multiétages et applications aux détecteurs de photons et à l'imagerie de rayonnements ionisants.
	Participation à des expériences à Fermilab (USA)
1985–1991	Introduction des chambres à avalanches lumineuses. Développement d'appareillage pour les recherches en biologie utilisant l'imagerie des rayons β (Centre Médical Universitaire de Genève)

Décorations
Croix de Guerre 39–45

Distinction
Prix Ricard de la Société de physique (1980)
Prix du Commissariat à l'énergie atomique de l'Académie des Sciences (1984)
Docteur Honoris Causa de l'Université de Genève (1980)
Académie des Sciences (France) (1985)
Foreign Associate of the National Academy of Sciences of the USA (1986).

ELECTRONIC IMAGING OF IONIZING RADIATION WITH LIMITED AVALANCHES IN GASES

Nobel Lecture, December 8, 1992

GEORGES CHARPAK

CERN, Geneva, Switzerland

Detecting and localizing radiation is the very basis of physicists' work in a variety of fields, especially nuclear or subnuclear physics.

Certain instruments have assumed special importance in the understanding of fundamental phenomena and have been milestones in the building up of modern theories. They make up a long list: the ionization chamber, the cloud chamber, Geiger-Müller counters, proportional counters, scintillation counters, semiconductor detectors, nuclear emulsions, bubble chambers, spark and streamer chambers, multiwire and drift chambers, various calorimeters designed for total absorption and then measurement of particle energy, Cherenkov or transition radiation counters designed to identify or select particles, and many other detectors, some very important examples of which are still being developed. However, some of this equipment has become obsolete as physicists' requirements have changed. Wire and drift-chambers, introduced in 1968, met the then requirements of physicists, whereas the properties of the most productive detectors available at the time, mainly bubble and spark chambers, were no longer capable of meeting those needs.

Multiwire chambers gave rise to further developments in the art of detectors, of which some are highly innovative. Most high-energy physics experiments make use of these methods, but their application has extended to widely differing fields such as biology, medicine, and industrial radiology.

Our study of multiwire proportional chambers, which began in 1967, was triggered by the problems with spark chambers which then faced us. The latter, introduced in 1959 by Fukui and Myamoto, beautifully supplemented the bubble chamber. Whereas the latter was still peerless in the quality of the information which it provided and from which one single exposure could on its own lead to an interesting discovery, the spark chamber gave a repetition rate more than 100 times higher. Moreover, as it had a memory of almost 1 μs, the instrument could be triggered only for events selected by faster auxiliary counters, making it possible to address the study of phenomena which occurred much more rarely in very high-energy interactions. Nevertheless, the need to store the information on photographic films led to a bottleneck: beyond a few million photographs per year or per experiment, the exposure analysis equipment was saturated.

Physicists therefore had to invent methods of reading the sparks which bypassed photographs. We introduced two new methods: one was based on the measurable delay of the signal produced by a spark in reaching the end of an electrode; the second, based on the division of the current produced by a spark in plane or wire electrodes, at the ends of which the current pulses are measured, made it possible to obtain the coordinates of the spark and hence of the particle by purely electronic means. This latter method was developed in several laboratories for the focal planes of spectrometers and we have ourselves thus performed experiments on the nuclear reactions induced by pions.

Other approaches, some of them better than ours, were developed simultaneously: sonic spark chambers and wire spark chambers which gave rise to very important developments. Nevertheless, the fact that it is impossible to trigger spark chambers at rates above about 100 times per second reduced their scope. In the minds of some physicists arose the idea of limiting the discharge produced from electrons released in a gas to a much lower level than that attained by the spark so as not to discharge the capacitance formed by the electrodes; the additional gain thus needed was to be obtained by means of electronic circuits.

In 1967, I undertook this step, armed with some experience acquired at the Collège de France from 1948, some ten years before I joined CERN. I built cylindrical single-wire proportional chambers and also demonstrated the possibility of making use of the light phenomena produced by an avalanche of electrons in a gas. This approach led to no practical method. It was greatly extended later on during work done at the University of Coimbra in Portugal by E. Policarpo. This work, from which I drew a great deal of inspiration, proved very valuable for the experience I gained and, more particularly, the understanding of the principles governing the multiplication of electrons in gases. It led me to an attempt to build chambers with avalanches made visible by short electric pulses. It resulted, in 1956, in the first detector with sparks following the trajectory of particles. It played no role in the introduction of spark chambers in particle physics.

Figure 1 shows the design selected in 1967 to study proportional multiwire structures.

A study of the electric fields shows that, in the region near the wire taken to a positive potential, where a limited-scale avalanche is to be produced, the electric field is the same as that prevailing near a wire tensioned in the axis of a cylindrical tube, as can be seen in figs. 2 and 3. With the parameters we chose, in a gas in common use in proportional counters, the average mean free path of an ionizing collision should, at atmospheric pressure, be about 1 μm (fig. 4). We might therefore expect a gain of about 10^5 for an avalanche extending over a distance close to the diameter of the wire, i.e. 20 μm.

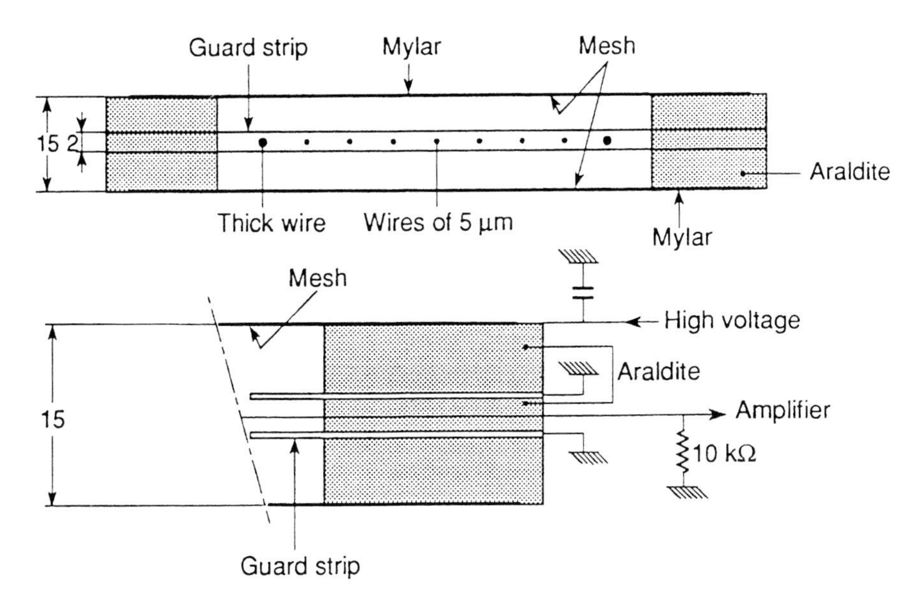

Fig. 1: A few construction details of multiwire chambers. The sensitives anode wires are separated by 2 mm from each other; their diameter is 20 μm. They are stretched between two cathode meshes, in a gas at atmospheric pressure. The edges of the planes are potted in Araldite, allowing only the high voltage to enter and only the pulses to leave to go to a 10 kΩ amplifier.

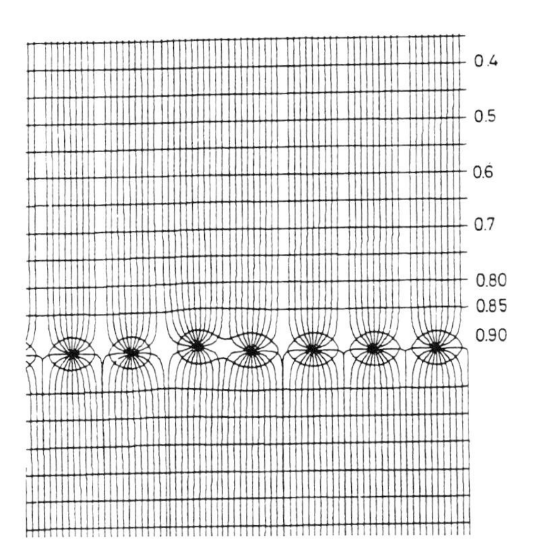

Fig. 2: Equipotentials and electric field lines in a multiwire proportional chamber. The effect of the slight shifting of one of the wires can be seen. It has no effect on the field close to the wire.

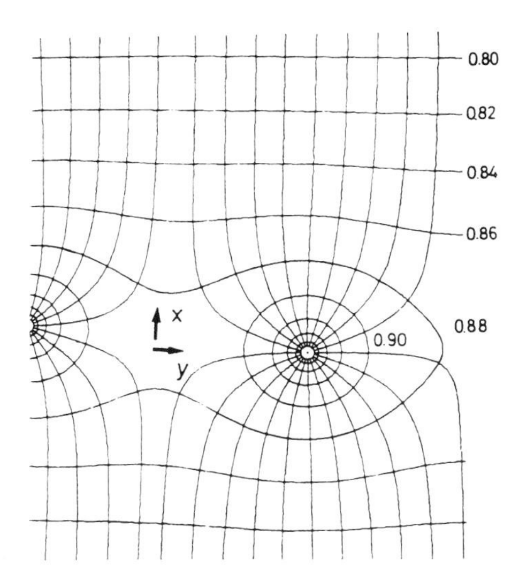

Fig. 3: Detail of fig. 2 showing the electric field around a wire (wire spacing 2 mm, diameter 20 μm).

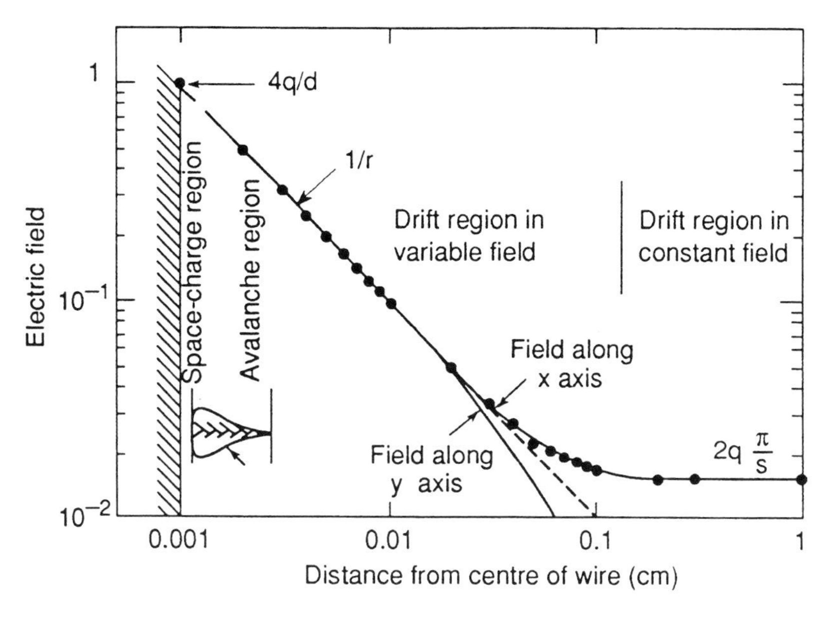

Fig 4: Variation in a proportional chamber of the electric field along the x axis perpendicular to the plane of the wires and centred on one of them, and in the y direction parallel to this plane. The various regions (electron, avalanche and drift space in a variable and then constant field) are shown.

There still remained one challenging problem, that of the capacitive coupling between the wires. The closer they were, the more likely it was that a pulse induced in one wire would be propagated on its neighbours. This was true for pulses produced by an external electric generator but untrue for the internal generator formed by the positive and negative ions separating under the effect of the electric field. There had, in the past, been examples of wire counters, especially in cosmic-ray experiments, where this fear of coupling led to the insulation of each of the positive amplifying wires by partitions or thick intermediate wires. We merely have to examine the pulse-generation mechanism in a proportional counter to see that, whatever the distance between the wires, the one which is the seat of an avalanche will develop a negative signal, whereas the neighbouring wires and, in general, all the neighbouring electrodes, develop a positive signal which is therefore easy to distinguish from the other.

Most of the electrons produced in the first microns in front of the wire pass through a very small proportion, ΔV, of the potential V applied between the wire and the cathode on their trajectory. The collected charge Q will produce on the wire taken to the potential V only a pulse of charge ΔQ, so that $V \cdot \Delta Q = Q \cdot \Delta V$. With our selected parameters, the time needed to collect the electron charge was a fraction of a nanosecond. However, the positive ions have to pass through the whole of the potential drop V and thus induce almost all of the charge pulse which develops as a function of time according to a law which reflects the considerable field close to the wire and the decreasing field far from it. The initial very fast increase of the pulse gave some people the illusion that the initial pulse observed was caused by the collection of the electrons of the avalanche. For a proportional counter with a radius of 1 cm, and a wire with a diameter of 20 μm, the electron contribution is only 1 %. Figure 5 shows the characteristic shape of the development of a pulse in a proportional chamber owing to the motion of the ions. Although the total time taken for ion collection is close to 500 μs in the example chosen, almost half of the signal develops within a time close to a thousandth of this value, which is very useful for fast detection.

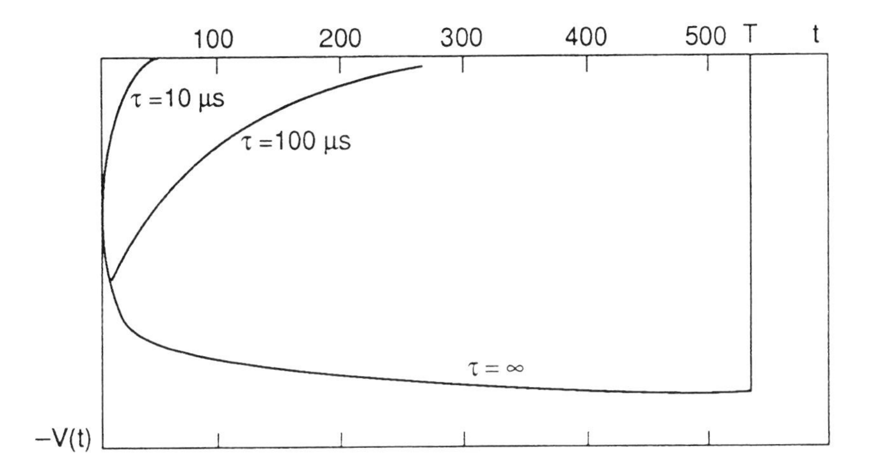

Fig. 5: Development in time *t* of a pulse in a proportional chamber; *T* is the total drift-time of the positive ions between the anode and the cathode. The effect of decreasing differentiation time constants τ is seen. The motion of the positive ions close to the wires produces a very fast-rising pulse .

The key phenomenon is this: whereas the induced signal is negative on the wire from which the positive ions are going away, it is positive on the neighbouring wires or electrodes.

This has two important consequences:

- It is a simple matter to locate the wire which is the seat of the avalanche, whatever the distance between the wires.
- The position of this avalanche along the wire can be obtained if the cathode is made of wires or strips perpendicular to the anode wire.

The distribution of the induced positive signals is then centred on the avalanche. Experience was to show that this observation was of prime importance to the imaging of neutral radiation, photons, or neutrons.

The time resolution of the detector depends on the distance between the wires. Proportional cylindrical chambers had been abandoned in particle-physics experiments as this resolution was poor: the time taken by the electrons released in the gas to reach the multiplication region near the wire is actually variable. It immediately appeared that the delay was easily measured in a multiwire chamber and that it gave precisely the distance of the ionizing particle from the wire.

This cleared the way for a class of detectors derived from wire chambers in which the detecting wires are very far apart and where the coordinate is calculated by measuring the drift-time of the electrons in the gas. From 1968 we showed that it was possible to obtain accuracies of the order of 100 μm with structures such as that of fig. 6, which provide a constant electric field over a long distance. In 1969, a group at Saclay began to build drift chambers with a migration length of 20 cm, while, in Heidelberg, studies were undertaken on chambers constructed, similarly to ordinary wire chambers, with a field wire fitted between the anode wires to repel the electrons. The drift chamber seemed to be the ideal instrument for large detectors; many were the groups which then began to fit considerable areas with these detectors, reaching, for instance, 5 m×5 m, making possible a precision of a few hundred microns, with electronics comprising a limited number of channels.

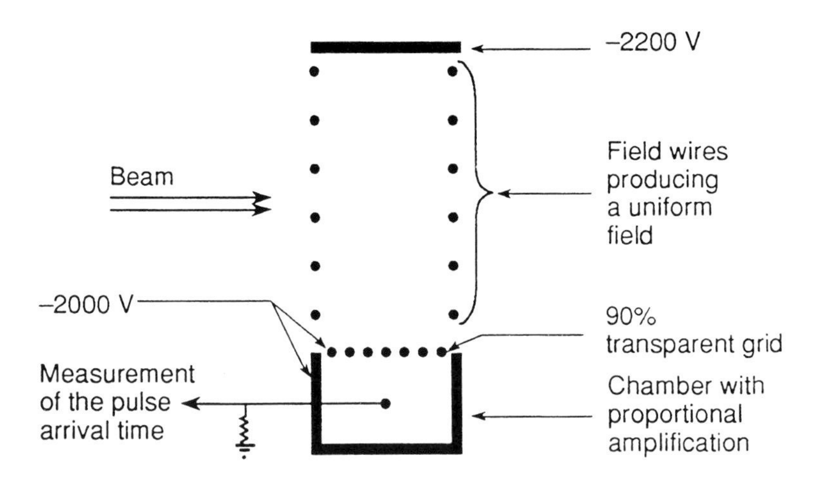

Fig. 6: Operating principle of a detector based on the drift-time of the electrons in a constant electric field (1968).

Starting in 1969, we stressed that the exploitation of electron drift in large volumes, combined with the measurement of the centroid of the avalanches induced in a wire chamber, was the way leading to threedimensional detectors. It was, however, D. Nygren who, by combining the effects of parallel magnetic and electric fields and solving formidable data acquisition problems, succeeded in creating an instrument which provides the finest images of the most complex configurations obtained in colliders; for this purpose the repetition rate must be low enough to accommodate an electron drift over long distances. In fact it was the demands of physics, differing very widely depending on the nature of experiments and accelerators, which dictated the structures of gas detectors, making use of the properties that we demonstrated in 1968.

The development of transistor electronics, however, made it possible to design systems requiring tens of thousands of channels. The advantage of wire over drift chambers was their capacity of accepting very high counting rates. The resolution time, of about 30 ns, and the possible counting rate of 10^5 pulses per second, made it feasible to tackle the study of rare phenomena which were beyond the reach of spark chambers as they needed very high counting rates.

The design study of a giant detector, the Split-Field Magnet (SFM) was launched at CERN in 1970 under the direction of A. Minten for an experiment at the Intersecting Storage Rings (ISR). This detector comprised 70,000 wires, some of them 2 m long. Another group under J. Steinberger undertook the construction of a detector designed to collect a large number of events violating charge parity conjugation in the study of kaon decay. A large number of difficult problems had to be solved to go from the 10 cm$\times$10 cm chamber to these large areas. The intensity of the pulses collected on the wires of a chamber was proportional to the energy deposited in the volume defined by the electric field lines ending at a wire. In a cylindrical counter, simple considerations showed that the logarithm of the gain was proportional to a factor of $f = V^{1/2}[(V^{1/2}/V_t^{1/2})-1]$, where V is the applied voltage and V_t the threshold voltage. The behaviour of the wire chamber is exactly the same as that of a cylindrical counter.

My group was then reinforced by F. Sauli, who, together with collaborators joining the various projects launched by us, contributed greatly to the success of many new detectors. We undertook a systematic study of the factors controlling the accuracy in drift chambers. We investigated the ultimate accuracy which can be reached in multiwire chambers by measuring the signals induced on the cathodes. We showed that the avalanches could extend to only a very limited extent around the wire and that by measuring the centroid of the induced signals it was possible to determine the azimuth of an avalanche. Our results and those of a few others, which started a systematic study of the multiwire structures, led to a generalization of the use of wire chambers and drift chambers with a considerable diversification of the detector's structure best adapted to the variety of situations encountered in particle physics.

Mixtures of up to four gases were found to reduce the cost of the necessary electronics, making it possible to obtain high saturated pulses independent of the energy deposited in the gas, and required less sensitive and expensive electronics than with the proportional system, as well as being sufficiently resistant to ageing effects.

It was demonstrated that these pulses were produced by a series of avalanches which stopped at the uniform low field far from the wire.

A systematic study of chambers filled at very low pressures, conducted in Israel by A. Breskin, was to show that the chambers operated at pressures as low as 1 Torr with astonishing time resolutions: the range of application of wire chambers was widening.

Whereas wire and drift chambers were essential in all the particle physics experiments, rapidly replacing spark chambers and, in some instances, reinforcing bubble chambers with structures outside them, many groups were making use of the new opportunities offered for imaging various types of ionizing radiation. The main applications appeared in the field of X-rays with an energy close to 10 keV. Under the inspiration of V. Perez-Mendez, chambers were built to study the structure of proteins by the imaging of X-rays diffracted by their crystals. Gas detectors had the drawback of being highly transparent to X-rays and the methods used to overcome this required compressed xenon.

We, on our side, tried to solve this problem by building a spherical drift chamber centred on the macromolecule crystal. The radial lines of the electric field eliminate any parallax and the electrons drifting over 15 cm are transferred into a 50 cm$\times$50 cm multiwire chamber in which the avalanches are measured to a precision of 0.5 mm. It is also found that the response is continuous in both dimensions, as the diffusion ensures an expansion of the cloud of ionization electrons which always covers two wires, making it possible to interpolate the position of the avalanche between them.

The apparatus has considerable advantages over photography in terms of data-acquisition rate and the signal-to-noise ratio. Used routinely with the X-ray beams produced by the synchrotron radiation from an electron storage ring at Orsay it gives a good crop of important results.

The imagination shown by various groups has made it possible also to extend the field of application of wire chambers for higher X- or γ-ray energies.

Thus, a Novosibirsk group has developed a chamber making it possible to X-ray the human body with a decrease in dosage over the most powerful equipment available on the market of at least a factor of 10.

A group at Schlumberger has produced a system for radiographing giant containers using X-rays of up to 5 MeV. A. Jeavons has constructed positron cameras capable of detecting 0.511 MeV γ-rays with a precision of about 1 mm. These cameras, which are not efficient enough for the method to be applied to nuclear medicine, made possible a remarkable advance in the field of solid-state physics. Finally, a firm is now marketing a camera for γ-rays, competing with the Anger camera, which is specially suitable for use on children. Its principle is based on a wire chamber filled with compressed xenon. It gives a considerable reduction in the doses administered and also increased precision.

These few examples are enough to show that we may be on the threshold of the general use of radiation detectors originally invented for particle physics. It will develop in line with the progress made in particle physics research laboratories.

An important stage in the widening of the field of application of gas detectors was made with the introduction, by J. Séguinot and T. Ypsilantis, of photosensitive vapours. They make it possible to locate photons in the far ultraviolet of an energy above about 5.3 eV with a precision of less than 1 mm. Major instruments designed to identify particles through Cherenkov radiation are now in use in some giant collider detectors. D. Anderson explored the opportunities provided by the detection of the photons emitted by scintillators. Figure 7 shows a 9 GeV photon spectrum obtained with a chamber containing a tetrakis(dimethylamine)ethylene (TMAE) vapour. Subsequent research also shows that it is possible to use condensed photocathodes compatible with gas amplification.

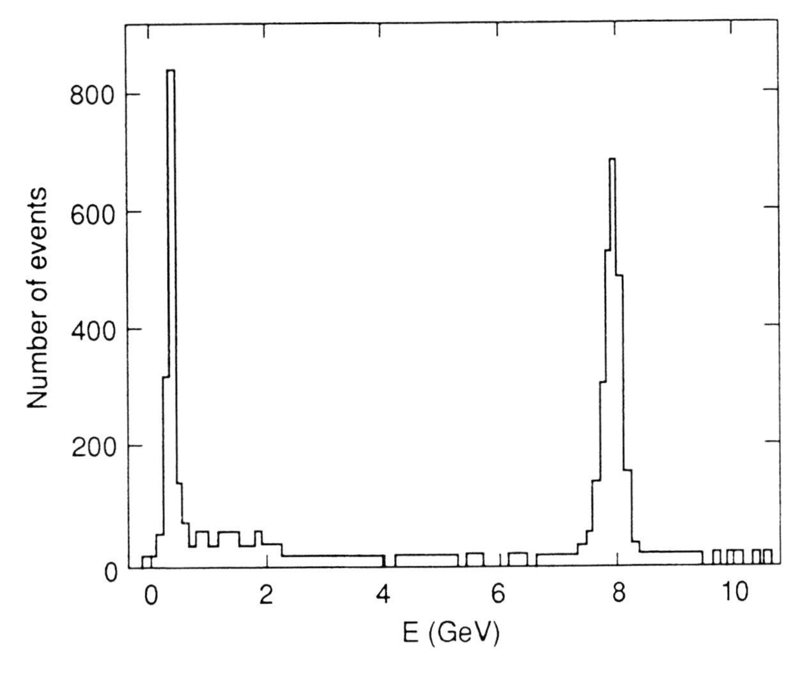

Fig. 7: Energy spectrum of 9 GeV particles (electrons and pions) with BaF_2 crystals coupled to wire chambers filled with a photosensitive gas (TMAE, see text)

As the intensity of accelerators increased, it was found that the wire chambers themselves could not cope with the expected counting rates. Beyond particle fluxes from 10^4 to 10^5 hits per second and per millimetre of wire, the space charge produced by the accumulation of positive ions offset the applied field and annulled the gain.

To overcome this problem I thought, together with F. Sauli, of separating the gas amplification into two stages: a gain in a preamplification structure followed by a partial extraction of the electrons into a drift region fitted with a control grid, and their transfer into a wire chamber which would be required only to amplify the electrons accepted by the grid (fig. 8). It is then possible to accept or reject an event with a time precision of about 30 ns and a delay defined by the electron drift-time. Together with S. Majewski we found a structure reaching this goal. It made it possible easily to amplify single photoelectrons released by an ultraviolet photon in a photosensitive gas. The drift region in fact largely eliminates the effect of the radiations emitted by excited atoms produced in an avalanche, which are responsible for secondary effects by ejecting electrons near the initial position of the photoelectrons. We have built large detectors of this kind in collaboration with Saclay to obtain images of Cherenkov rings in an experiment performed at the Fermi Laboratory Tevatron in the United States. This development proved most fruitful, however, in a special biological field of application. There are several research fields in which it is necessary to obtain the image of the distribution of molecules marked with radioactive elements. Several commercial firms have tried to make use of wire chambers for this imaging purpose. The main difficulty lies in the generally considerable distance which the electrons from the radioactive bodies may have to travel in the gases. We have observed that an amplifying structure based on parallel grids did not suffer from this defect. Multiplying an electron by a Townsend avalanche, in fact, exponentially helps the ionization electrons released in the gas near the cathode, which is the entry window. Precisions of the order of 0.3 mm have thus been obtained for β radiation emitted by phosphorus-32, for instance.

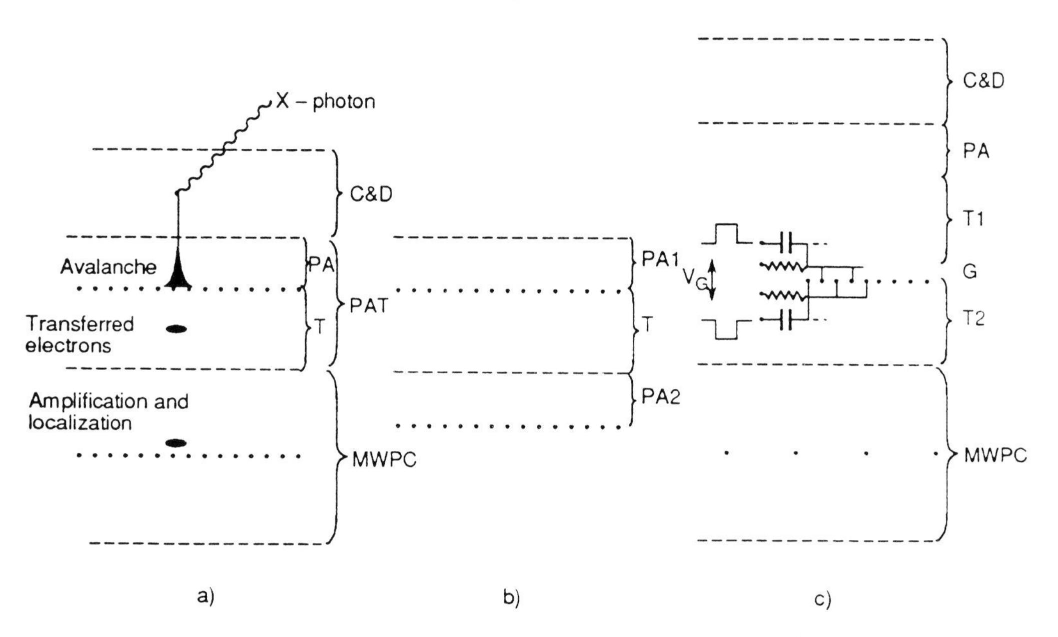

Fig. 8: Various multistage structures. (a) A preamplification (PA) and transfer (T) component is followed by a multiwire proportional chamber (MWPC); the conversion and drift space (C and D) makes it possible to inject a constant charge (X-rays etc.) into the amplification region. (b) Two parallel-plate units with charge transfer from PA1 to PA2. The best time resolutions were obtained with this configuration. (c) A multistage chamber with a gate: the wires of grid G are at alternating potentials of ± 70 V. The electrons are transmitted from component PA + T_1 to MWPC and they are transferred when the wires are brought to the same potential by short pulses of 30 ns, making it possible to select rare events in a high rate environment.

We then took another step by coupling such a multiplication region with a method of reading out the position of the avalanches, making use of the light emitted. We had developed this method in order to study very rare phenomena such as double β decay or collisions between gas atoms and certain types of hidden matter. We thought that only the redundancy provided by photographic methods could characterize a rare event. A British group was also independently tackling this method for imaging Cherenkov photons. They, too, used the multistage chambers which we had introduced and reached the same conclusion: with suitable vapours it was possible to obtain a sufficiently plentiful emission of photons to detect the light emitted by the avalanches with simple optics.

The great advantage of this method lies in the read-out which it permits in the use of hundreds of thousands of channels now making up the CCD (charge-coupling device), which is the basic component of video cameras. Moreover, the fact that the image of a light avalanche covers several pixels makes it possible to interpolate the centre of the avalanche and multiply the number of real channels by almost 10. We were thus able to make an instrument for visualizing the distribution of the electrons emitted by tritium in a slice of rat's kidney, making it possible to observe details of about 100 μm. The fact that the same data (fig. 9) were obtained in a single day instead of the three months previously required by the photographic method aroused immediate interest among certain biologists and gave rise to a development which is still under way.

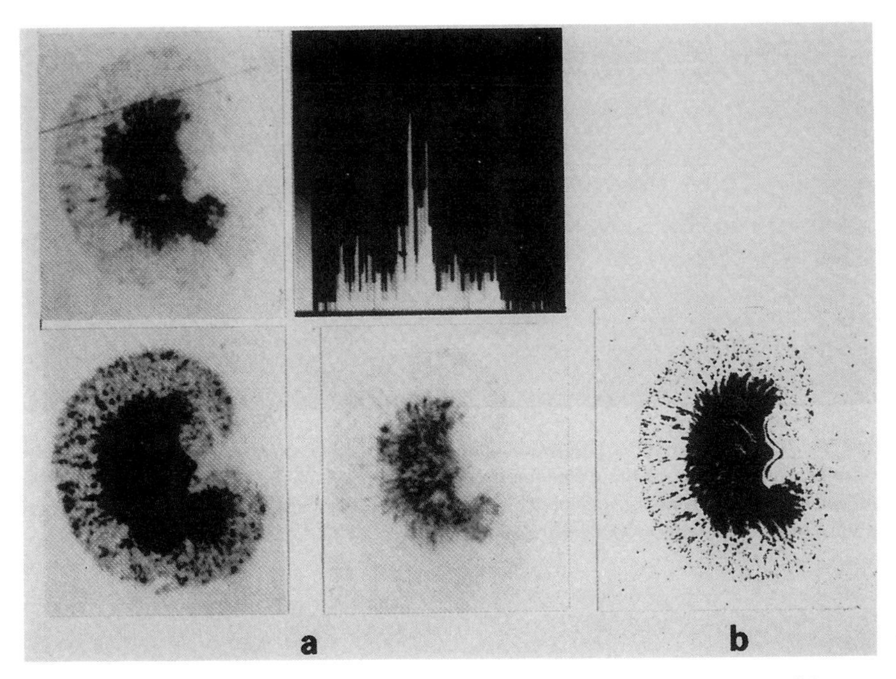

Fig. 9: Images of samples taken from a tritium-marked rat's kidney. The four parts of diagram (a) were obtained with a gas detector: the window, centred at three different levels, has a grey scale of 256. The intensity spectrum is taken along the oblique line marked on the first view with 40 μm pixels in the sample plane. The gas mixture is xenon with 2.5 % triethylamine. The details of the renal channels, measuring 50 μm, can be seen. This image may be obtained in 20 hours. (b) Autoradiograph showing the adjacent slice of rat's kidney to which a photographic emulsion was applied for 3 months.

Our latest experiment with gas detectors comprising parallel-grid structures helped convince us that the controlled multiplication of avalanches in gases, in structures as diverse as wires or parallel surfaces, by selecting photocathodes compatible with gas amplification is still a source of fruitful developments: in many fields this multiplication may still result in progress in all fields where radiation must be imaged, from ultraviolet photons to γ-rays and the highest-energy particles. At the current stage of high-energy physics, however, simply making use of the location of free electrons near the wires of proportional chambers and the drift-time of the electrons provides an image of configurations rivalling in complexity those provided by bubble chambers. This is shown in fig. 10, the image of an event generated in the ALEPH detector installed at one of the intersections of LEP, the large e^+e^- collider operated at CERN.

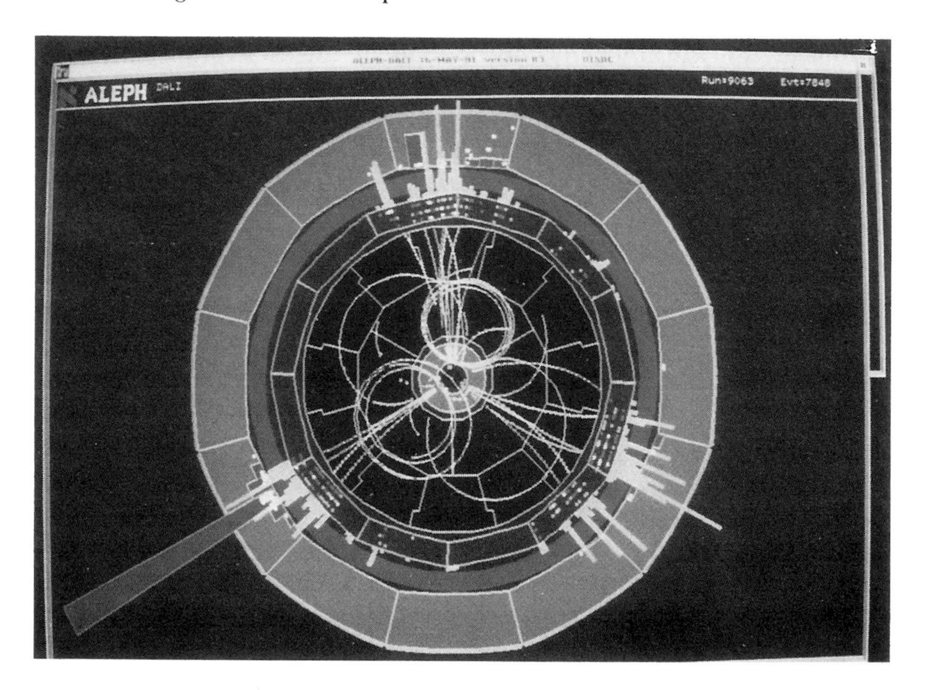

Fig. 10: Image of an e^+e^- collision obtained at the ALEPH experiment at LEP using an instrument making use of the drift-time in a large volume and the read-out of coordinates projected in a wire chamber. Auxiliary outside detectors provide the information on the energy of the particles, the trajectory of which was displayed.

BIBLIOGRAPHY

A very comprehensive historical study with most references corresponding to the period covered by this artide is to be found in

I. Gambaro, The development of electron detectors at CERN (1966–late 1970s), *CERN History Study No. CHS-39,* January 1992.

Our first article that described the properties of multiwire structures is

G. Charpak, R. Bouclier, T. Bressani, J. Favier and C. Zupancic, The use of multiwire proportional chambers to select and localize charged particles, *Nuclear Instruments and Methods* 62 (1968) 202–26.

It was followed by several articles, one of them giving a description of the progress we made during our first year of studies of the wire chambers and the drift chambers:

G. Charpak, D. Rahm and H. Steiner, Some developments in the operation of multiwire proportional chambers, *Nuclear Instruments and Methods* 80 (1970) 13–35.

We refer the reader to highly detailed works such as

F. Sauli, Principles of operation of multiwire proportional and drift chambers, *CERN* 77-09 (1977).

P. Rice-Evans, *Spark, streamer, proportional and drift chambers* (Richelieu, London, 1974),

T. Ferbel, *Techniques and concepts of high-energy physics* (Plenum Press, New York, 1987),

and to a book giving a good analysis of the development of detectors in relation to the evolution of particle physics:

F. Close, M. Marten and C. Sutton, *The particle explosion* (Oxford University Press, 1987).

Physics 1993

RUSSELL A. HULSE and JOSEPH H. TAYLOR, JR.

for the discovery of a new type of pulsar, a discovery that has opened up new possibilities for the study of gravitation

THE NOBEL PRIZE IN PHYSICS

Speech by Professor Carl Nordling of the Royal Swedish Academy of Sciences.
Translation from the Swedish text.

Your Majesties, Your Royal Highnesses, Ladies and Gentlemen,

This year's Nobel Prize in Physics has been awarded to Russell Hulse and Joseph Taylor for the discovery of a new type of pulsar, a discovery that has had a great impact on gravitational physics.

When a star dies and its light fades away, it may be transformed into a pulsar. It then vanishes from the visible firmament. We no longer see it. But it is there, sending out radio signals instead of light, and in its new guise it has taken on remarkable characteristics.

As an astronomical object it is now tiny, only about ten kilometers in diameter. It consists entirely of nuclear matter, chiefly neutrons. Its density is extremely high. One pinhead of matter in such a pulsar would weigh hundreds of thousands of tons. It rotates at enormous speed, perhaps approaching a thousand revolutions per second. It continuously emits a radio signal in two beams that sweep across space, resembling the light beam from a lighthouse.

A terrestrial radio receiving antenna aimed at this transmitter picks up a signal that pulsates at the same frequency as the extinguished star rotates. Its frequency is very stable, fully comparable with that of terrestrial atomic clocks. This pulsating radio signal was the origin of the name "pulsar."

During a series of observations employing refined techniques to study the occurrence of pulsars, Joseph Taylor and his doctoral student Russell Hulse made the discovery that is being rewarded with this year's Nobel Prize. At a position in the sky with the celestial coordinates 1913+16, they found a new pulsar. In itself, this was not a remarkable discovery, since many new pulsars had been identified in the course of their work.

But this object behaved differently from previously known pulsars. The time between its radio pulses—59 milliseconds—was not constant, but showed periodic changes. The pulsar was being subjected to some kind of disturbance. Hulse and Taylor supplied an explanation for this phenomenon that was simple, yet stirred the imagination: Their pulsar had an invisible companion!

These two objects orbit each other, with the pulsar sometimes moving toward the earth and the radio antenna, sometimes away from them. When the pulsar moves toward the earth, the antenna picks up a higher frequency signal. When the pulsar moves away from earth, the antenna receives a lower frequency signal. This phenomenon, called the Doppler effect, occurs in many contexts. We experience it every time we hear an ambulance siren changing its audible frequency as the ambulance first approaches, then moves away from us.

This effect was what revealed the presence of the invisible companion—invisible both to the eye and to the radio receiver with its 300 meter diameter antenna dish. This companion is probably also a neutron star. Perhaps it is a pulsar that emits two beams of radio frequency radiation into space, although neither of them sweeps over our little planet. But perhaps they sweep over some other planet. Perhaps a radio astronomer in a distant civilization is sitting right now, recording these pulses and pondering why "his" pulsar is demonstrating certain irregularities in its radio signals.

Hulse and Taylor found that the pulsar moved at a speed of up to 300 km per second in its whirlwind dance around its companion. This is a high speed, ten times higher than the speed at which the earth travels in its orbit around the sun. Hulse and Taylor realized that their binary pulsar thereby provided a unique opportunity to observe effects that Einstein had predicted sixty or seventy years earlier in his theories of relativity. New hope was awakened among relativity theorists throughout the world. Now they had an object tens of thousands of times more favourable than the planet Mercury, the classic object for testing the general theory of relativity and competing theories.

One of the most fascinating predictions of relativity theory is that massive objects in vehement motion emit a new kind of radiation, known as gravitational radiation. This phenomenon is also described as a wave motion, as ripples in the curvature of space-time, and we speak of "gravitational waves."

No one has yet succeeded in recording a gravitational wave in a terrestrial or extraterrestrial receiver, but the Hulse-Taylor pulsar has convinced us that this type of radiation actually exists. This is because the orbiting period of the pulsar around its companion gradually diminishes with time—extremely little, but in exactly the way the general theory of relativity predicts, as a result of the emission of gravitational waves. This triumph puts the Hulse-Taylor pulsar in a class by itself as a laboratory for gravitational physics.

Dr. Hulse, Professor Taylor,

You have been awarded the 1993 Nobel Prize in Physics for your discovery of the first binary pulsar, PSR 1913+16, a discovery which has had a great impact on gravitational physics. It is my privilege to convey to you the heartiest congratulations of the Royal Swedish Academy of Sciences, and I now ask you to receive the Prize from the hands of His Majesty the King.

Russell A. Hulse

RUSSELL ALAN HULSE

I was born November 28, 1950 in New York City, the son of Alan and Betty Joan Hulse. My parents tell me that I quickly showed an unusual level of curiosity about the world around me as a child, and that this transformed itself into an interest in science at a very early age. For my part, I certainly recall that science was a defining part of my approach to life for as far back as I can remember. My parents fostered and supported this interest, and I thank them very much for being my first and, by far, most uncritically supportive funding agency. I ran through a seemingly endless series of interests involving chemistry sets, mechanical engineering construction sets, biology dissection kits, butterfly collecting, photography, telescopes, electronics and many other things over the years.

The door to a whole range of new experiences opened for me when my father started building a summer house on land given to us by my Aunt Helen in Cuddebackville, New York, about two hours northwest of the city. Eventually, this became a year-round house for my grandparents when they retired and it is where my parents live now that they are retired. I remember spending weekends and summers helping my father put in place walls, rafters, siding and everything else that goes into a house. Among other things, it produced an early familiarity with tools and a do-it-yourself approach which has stood me in good stead over the years. My parents' friends and relatives were apparently not too sure that I should have been given such freedom to work with power tools at an early age, but fortunately I came through the experience with all of my fingers intact. Cuddebackville was also important to me as a place where a city kid could see nature, and as a practical place to work on my bigger projects.

My parents not only supported my interests at home but also suffered along with me (and, most likely, much more than me) when some of my first experiences with school proved to be less than positive. Though I had some elementary school teachers with whom I got along well, there were some real problems with others who found me and my intense interest in science difficult to understand and deal with.

Entering the Bronx High School of Science in 1963 was thus very important to me as it was there that I found myself in a school environment which explicitly emphasized what I found most interesting in life. Yet, as in the years before and after, while schoolwork was an important job to be done my interests in science tended to be expressed most clearly by my home projects. My biggest home project while at Bronx Science was building an amateur radio telescope up at my parents' house in Cuddebackville. I particularly enjoyed building antennas of various types, relying on an amateur radio antenna design book as a guide. The electronics were an odd mix of old television parts, military surplus power supplies, receivers and the like combined with other components I built myself. Unfortunately, the telescope never did work particularly well in terms of detecting radio sources (a little outside technical advice probably would have made a big difference in there somewhere), but I did enjoy myself and I learned a lot in the process.

At the end of high school, I had my first big career decision to make. While I had by then begun to focus more on physics and astronomy amongst the sciences, I also enjoyed designing and building electronic equipment. This lead me to consider electrical engineering as well but, in the end, I decided that a degree in physics was probably the best fit to my interests.

My college choices were limited by the fact that paying for college would have placed an inordinate financial burden on my parents. Fortunately, I was admitted to Cooper Union, a tuition-free college in lower Manhattan. From 1966 to 1970, I lived at home in the Bronx with my parents and commuted to Cooper each day on the New York subway system. Along with the usual course work, Cooper provided me with my first experience with a new interest, computers. Cooper had an IBM 1620 available for the students to use and, while there were no courses on programming it, there were the instruction manuals. The first project that I selected by way of teaching myself FORTRAN was to use the computer to do orbit simulations, perhaps an early omen of things to come.

After receiving my bachelor's degree in physics from Cooper Union in 1970, I started graduate school at The University of Massachusetts in Amherst. While I knew that I eventually wanted to do my thesis research in astronomy, preferably radio astronomy, I once again leaned towards a broader background and decided to get my doctorate in physics rather than astronomy. I went to UMass not only because its graduate program offered this type of flexibility, but also because it was located not too far from New York in a rather beautiful part of rural western Massachusetts.

The five years I spent in Amherst are some of those which I remember most clearly from my past. Graduate school was an entirely new environment, with new experiences and challenges. The demands were such that, for the first time, I focused almost exclusively on my academic career, with my other outside interests tempered by the demands of the moment.

After passing my Ph.D. qualifying examinations, I turned to finding a thesis project. This represented at long last a convergence of my outside and career interests, as I finally started working in radio astronomy again, now as a career rather than as a hobby. The rest of that story is told in my Nobel lecture.

After completing my Ph.D. in 1975, I had a post-doctoral appointment at the National Radio Astronomy Observatory in Charlottesville, Virginia from 1975 to 1977. While I still enjoyed doing pulsar radio astronomy, from the moment I arrived at NRAO I was increasingly preoccupied with the lack of long-term career prospects in astronomy. While I had some confidence that I could find another position of some sort after NRAO, it was not at all clear to me when, where, and how I would be able to settle down with some reasonable expectation of stability in my career. I certainly knew of astronomers who had been obliged to roam from place to place for many years and the potential for such repeated major dislocations in my personal life was more than I could quite tolerate. In particular, I had the classic problem of how a two-career couple could stay in reasonable geographical proximity, since my friend, Jeanne Kuhlman, was then doing her graduate work at the University of Pennsylvania. I therefore decided to try falling back on my broader interests and my physics Ph.D., exercising the option which I had left myself when I started at UMass.

While even with this broader view not many good career opportunities seemed available, I did discover from an advertisement in Physics Today that the Princeton University Plasma Physics Laboratory (PPPL) was hiring. Not only did controlled fusion seem an interesting and diverse field, but the lab was located in Princeton, not too far from Jeanne in Philadelphia.

After interviewing at PPPL, I was offered a position with the plasma modeling group, based on my physics and computer background. Starting at the lab in 1977, my first task was developing new computer codes modeling the behavior of impurity ions in the high temperature plasmas of the controlled thermonuclear fusion devices at PPPL. I had never really done computer modeling before and the art and science of computer modeling is one of the most valuable things which I have learned in the 16 years which I have now been at the lab.

The multi-species impurity transport code which ultimately grew out of this initial work at PPPL is still in use to this day. It models the behavior of the different charge states of an impurity element under the combined influences of atomic and transport processes in the plasma. I oriented my development of this code very much towards its practical use by spectroscopists and other experimentalists in interpreting their data and one of my greatest satisfactions has been that this code has become widely used over the years both at PPPL as well as at other fusion laboratories. My own research with this code included determining transport coefficients for impurity ions by modeling spectroscopic observations of their behavior following their injection into the plasma. In connection with modeling impurity behavior, I also worked on investigating the atomic processes themselves, for example, by helping to elucidate the importance of charge exchange reactions between neutral hydrogen and highly charged ions as an important recombination process for impurities in fusion plasmas. In a rather different sort of contribution, I more recently developed a computer data format which has been adopted by the International Atomic Energy Agency as a standard for the compilation and interchange of atomic data for fusion applications.

While I am still involved in supporting this impurity transport modeling code at PPPL, my more active area of work in the past few years has been modeling the transport of electrons in the plasma as revealed by pellet injection experiments. The pellets involved here are pellets of solid hydrogen, injected at high velocity into the plasma. The relaxation of the plasma electron density profile after a pellet has deposited its mass inside the plasma provides an important way of observing plasma transport in action. For this work, I wrote an electron particle transport code which focused on modeling the experimentally observed density profile evolutions using theoretically motivated, highly non-linear forms for the particle diffusion coefficients.

In another recent new direction, I have been working to establish a new effort at PPPL in advanced computer modeling environments. The objective of this research is the development of novel approaches to creating modular computer codes which will make it much easier to develop and apply computer models to an extended range of applications in research, industry and education. I have been pursuing this work in the context of cooperative research and development agreements with an industrial partner, taking advantage of this new type of collaborative arrangement recently made possible between government sponsored research laboratories and the private sector.

By now, it is surely clear that my interest in science has never been so much a matter of pursuing a career *per se*, but rather an expression of my personal fascination with knowing "How the World Works", especially as it could be understood directly with hands-on experience. This central motivation has been expressed over the years not only in my career but also in a wide range of hobbies. Notable amongst these "hobbies" have always been interests in various areas of science beyond whatever I was professionally employed in at any given time. For example, I have most recently been considering that much of what I have found so interesting about both the natural and man-made world has involved how individual, often autonomous, elements combine to make a functioning whole, either by design or by self-organization. I have thus started to be interested in various aspects of the new so-called "sciences of complexity", especially as they can be explored using computer modeling.

My list of more traditional hobbies and recreational activities has also changed over time. Many activities which I formerly enjoyed, such as amateur radio and woodworking, have been eventually dropped simply because I realized that I did not have enough time and energy to pursue everything I might enjoy doing. A current list of my activities would include nature photography, bird watching (and observing the beauty and drama of nature in general), target shooting, listening to music, canoeing, cross-country skiing, and other outdoor activities.

I do not pretend to be anything like an accomplished expert in all of the many things that I have ever been or am presently involved in doing. My most fundamental urge has always been just to spend time on what I found the most interesting, trying of course to match this up somehow with the more practical demands of life and a career. In this sense I have come to realize that at times I must not have always been the easiest person to have had as a student, or as an employee, and I therefore appreciate the efforts of those who helped me to accommodate myself to these practical demands, or often, who worked to help accommodate the practical demands to me.

I would like to close on the thought that some of the most enjoyable moments of my life have always involved sharing my various interests with those others who understood them (and me) the best. Thus special thanks go to my parents, to Jeanne Kuhlman, and to all of the good friends that I have had over the years.

THE DISCOVERY OF THE BINARY PULSAR

Nobel Lecture, December 8, 1993

by

RUSSELL A. HULSE

Princeton University, Plasma Physics Laboratory, Princeton, NJ 08543, USA

Exactly 20 years ago today, on December 8, 1973, I was at the Arecibo Observatory in Puerto Rico recording in my notebook the confirming observation of the first pulsar discovered by the search which formed the basis for my Ph.D. thesis. As excited as I am sure I was at that point in time, I certainly had no idea of what lay in store for me in the months ahead, a path which would ultimately lead me here today.

I would like to take you along on a scientific adventure, a story of intense preparation, long hours, serendipity, and a certain level of compulsive behavior that tries to make sense out of everything that one observes. The remarkable and unexpected result of this detective story was a discovery which is still yielding fascinating scientific results to this day, nearly 20 years later, as Professor Taylor will describe for you in his lecture. I hope that by sharing this story with you, you will be able to join me in reliving the challenges and excitement of this adventure, and that we will all be rewarded with some personal insights as to the process of scientific discovery and the nature of science as a human endeavor.

PULSARS

Pulsars were first discovered in 1967 by Antony Hewish and Jocelyn Bell at Cambridge University, work for which a Nobel Prize was awarded in 1974. At the time, they were engaged in a study of the rapid fluctuations of signals from astrophysical radio sources known as scintillations. They were certainly not expecting to discover an entirely new class of astrophysical objects, just as we were certainly not expecting to discover an astrophysical laboratory for testing general relativity when we started our pulsar search at Arecibo several years later. Pulsars have indeed proven to be remarkable objects, not the least for having yielded two exciting scientific stories which began with serendipity and ended with a Nobel Prize.

The underlying nature of pulsars was initially the subject of intense debate. We now know that these remarkable sources of regularly pulsed radio emission are in fact rapidly spinning neutron stars. A sketch of a

PPPL#93X0345A

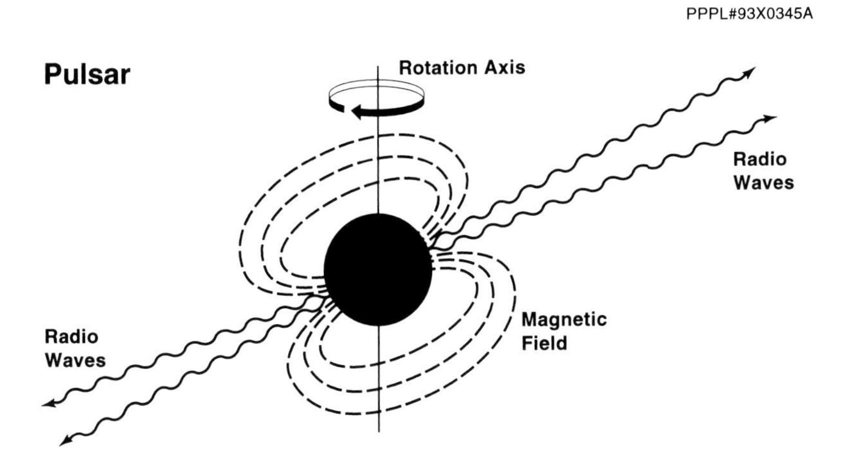

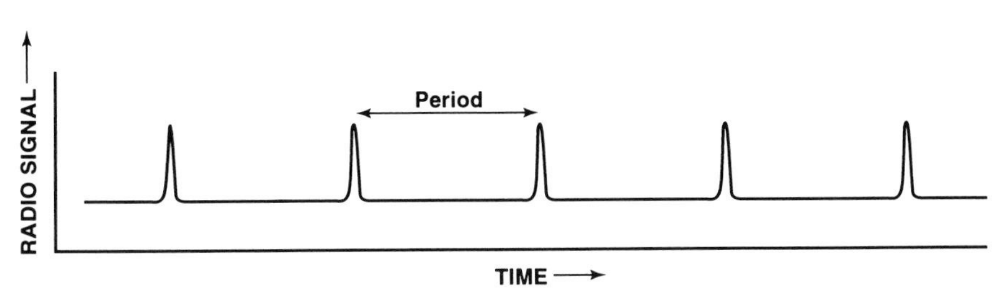

Fig. 1: A conceptual sketch of a pulsar, showing a rapidly rotating neutron star emitting narrow beams of radio waves from the polar regions of its embedded magnetic field. Also shown is a sketch of the periodic radio signal produced by a pulsar as seen by a radio telescope at the earth.

pulsar and its associated pulses as seen by a radio telescope is shown in Figure 1.

Neutron stars have roughly the mass of our sun compressed into an object only ~ 10 km in radius. Since the radius of our sun is about 7.0×10^5 km, one can immediately see that we are dealing with extraordinarily dense objects indeed. The intense gravitational field of a neutron star is sufficient to crush the electrons and nuclei of ordinary atoms into matter consisting primarily of neutrons. In a sense, a neutron star is a giant atomic nucleus, held together by gravitational rather than nuclear forces.

Neutron stars, and hence pulsars, are formed in supernovae, the spectacular final stage of stellar evolution for stars of many times the sun's mass. When the nuclear fuel of such a massive star is finally consumed, the resulting collapse of the stellar core produces a neutron star together with a cataclysmic explosion which expels the remaining stellar envelope.

The origin of pulsars in supernova explosions explains why in 1973 at the start of my thesis it was not considered surprising that all the pulsars detected to that point were solitary stars, despite the fact that a high fraction of normal stars are found in orbiting multiple star systems. It seemed reasonable to expect that the immense mass loss of a star undergoing a supernova would tend to disrupt the orbit of any companion star at the time that a pulsar was formed. The discovery of the first binary pulsar produced an immediate addendum to this conventional wisdom, to the view that while disruption is quite likely, it is not quite obligatory.

The conservation of angular momentum and magnetic field during the stellar collapse which forms a neutron star leads to the fast rotation of pulsars along with an intense magnetic field on the order of $B \sim 10^{12}$ Gauss. This field is sketched in Figure 1 as predominantly dipolar, although actually it is inferred to possess some fine-scale structure as we will discuss in a moment. The combination of fast pulsar rotation with this high magnetic field gives rise to the very narrow beams of radio waves shown emitted from the region of the magnetic poles. If one (or, occasionally, both) of these beams happen to be aligned such that they sweep across the earth, we see them as the characteristic pulses from a pulsar. The effect is very similar to that of a lighthouse, where a continuously generated but rotating narrow beam of light is seen as a regular, pulsed flash by an outside observer each time the rotating beacon points in their direction.

Synchronously averaging together a few hundred individual pulsar pulses by folding the pulse time series shown in Figure 1 modulo the pulsar period produces an average pulse profile with a good signal-to-noise ratio. This average or integrated profile has a highly stable shape unique to each individual pulsar, just like a human fingerprint. They range in form from very simple single peaks to complex forms with double or multiple peaks, reflecting details of the magnetic field geometry of each individual pulsar. The widths of these profiles are characteristically narrow compared with the pulse period, having a duty cycle typically of the order of a few percent for most pulsars.

For the purposes of the pulsar observations which we will be discussing here, it is this average pulse that we will be referring to as the fundamental unit of pulsar observation. As an aside, however, it is interesting to note that the reproducibility of the average pulse shape for a given pulsar contrasts with a rich variety of quasi-random and regular variation in the individual pulses; they are not simply invariant copies of the average profile. In this sense, Figure 1 is an oversimplification in tending to show all the individual pulses as being the same. This stability of an average pulse contrasting with the pulse-to-pulse variation is understood by relating the average pulse shape (e.g., pulsar beam shape) to the fixed magnetic field geometry, while local smaller scale emission events moving within the constraints of this magnetic field structure generate the individual pulses.

Precision measurements of the intrinsic pulsation periods of pulsars as defined by the arrival times of these average pulses show that pulsars are extraordinarily precise clocks. This stability is readily understood as arising from the identification of the observed pulsar period with the highly stable rotation of a rapidly spinning, extremely compact $\backsim 1$ solar mass object.

That pulsars are such precise clocks is in fact a central part of the binary pulsar story. In his lecture, Professor Taylor will describe pulsar timing methods in some detail. Suffice it for me to note for now that for PSR 1913+16, the binary pulsar we are going to focus on today, measurements of the pulsar period are now carried to 14 significant places, rivaling the accuracy of the most accurate atomic clocks. So you can certainly use a pulsar to set your watch without any hesitation!

SEARCHING FOR PULSARS

When I was approached by Joe Taylor at the University of Massachusetts to see if I was interested in doing a pulsar search for my thesis, it did not take too long for me to agree. Such a project combined physics, radio astronomy, and computers—a perfect combination of three different subjects all of which I found particularly interesting.

By 1973, pulsar searching *per se* was not by any means a novel project. Since the first discovery of pulsars in 1967, there had been many previous searches, and about 100 pulsars were already known at the time I started my thesis work. However, the telescopes and analysis methods used for these earlier searches had varied quite widely, and while many were successful, there seemed to be room for a new, high sensitivity search for new pulsars.

The motivations for such a new search emphasizing high sensitivity and a minimum of selection effects were several. First, a large sample of new pulsars detected by such a search would provide more complete statistics on pulsar periods, period derivatives, pulse characteristics, distribution in our galaxy and the like, together with correlations between these properties. Also, as pulsar radio signals travel to us, they are slowed, scattered and have their polarization changed in measurable ways which provide unique information on the properties of interstellar space in our galaxy. Thus finding a larger sample of new pulsars, especially ones further away in our galaxy, would be very valuable for these types of studies as well. In addition to statistical studies of the pulsar population, the search might also hope to uncover individual pulsars of unique interest, such as ones with very short periods. Indeed, since short period pulsars and very distant pulsars were precisely the types of pulsars which were the most difficult to detect for technical reasons, these most interesting of pulsars were precisely those the most strongly discriminated against in previous work. We saw an opportunity to improve on that situation with a comprehensive new computer-based attack on the pulsar search problem.

(And, yes, Joe's proposal to the National Science Foundation seeking funding for a computer for this search indeed pointed out, amongst these other motivations, that the discovery of "even one example of a pulsar in a binary system" would be quite valuable as it "could yield the pulsar mass". But we certainly had not been exactly counting on such a discovery, nor could we possibly have envisioned all that such a discovery might lead to !)

THE ARECIBO TELESCOPE

Since pulsars are relatively weak radio sources which must be observed at high time resolution in order to resolve their individual pulses, large radio telescopes are needed to collect as much signal as possible. Pulsars also have very steep radio spectra, meaning that they are strongest at low radio frequencies and are most often observed in the 100MHz to 1000MHz range, frequencies similar to those used for television broadcasts. Hence, the first requirement for an ambitious new high-sensitivity pulsar search was to use the largest radio telescope available for low-frequency pulsar work. The answer to that requirement was simple: Arecibo.

At 1000' in diameter, the Arecibo radio telescope is the largest single-element radio telescope in the world. An aerial view of the Arecibo telescope is shown in Figure 2. In order to support such a large reflector, the telescope was built in a natural bowl-shaped valley in the northwest corner of the island of Puerto Rico, several miles inland from the coastal town from which it takes its name.

Despite its fixed reflector pointing directly overhead, the Arecibo telescope is steerable to 20 degrees away from the zenith via the use of movable feed antennas. These antennas are suspended from a platform constructed of steel girders suspended by cables some 426 feet above the surface of the reflector. Using the 96 foot long feed antenna operating at 430MHz together with the Observatory's excellent low-noise receivers, Arecibo provided

Fig. 2: The 1000' diameter Arecibo radio telescope, operated by Cornell University for the National Science Foundation as part of the National Astronomy and Ionospheric Center (NAIC).

just the capabilities needed for our pulsar search. The limited steering ability of the antenna was an acceptable tradeoff for obtaining this extremely high sensitivity, although it did have some interesting consequences in terms of the binary pulsar discovery, as we shall see later.

PARAMETER SPACE, THE FINAL FRONTIER

In order to detect a signal of any kind with optimal sensitivity, we must completely characterize its properties and then devise a "matched filter" which will take full advantage of these known properties. From this point of view, the three critical parameters of a pulsar's signal for a search are its dispersion, period, and pulse width, as illustrated in Figure 3. Dispersion refers to the fact that due to its propagation through the free electron density in the interstellar medium, the pulses arrive at the earth successively delayed at lower frequencies. The existence of a characteristic period for

PPPL#93X0374

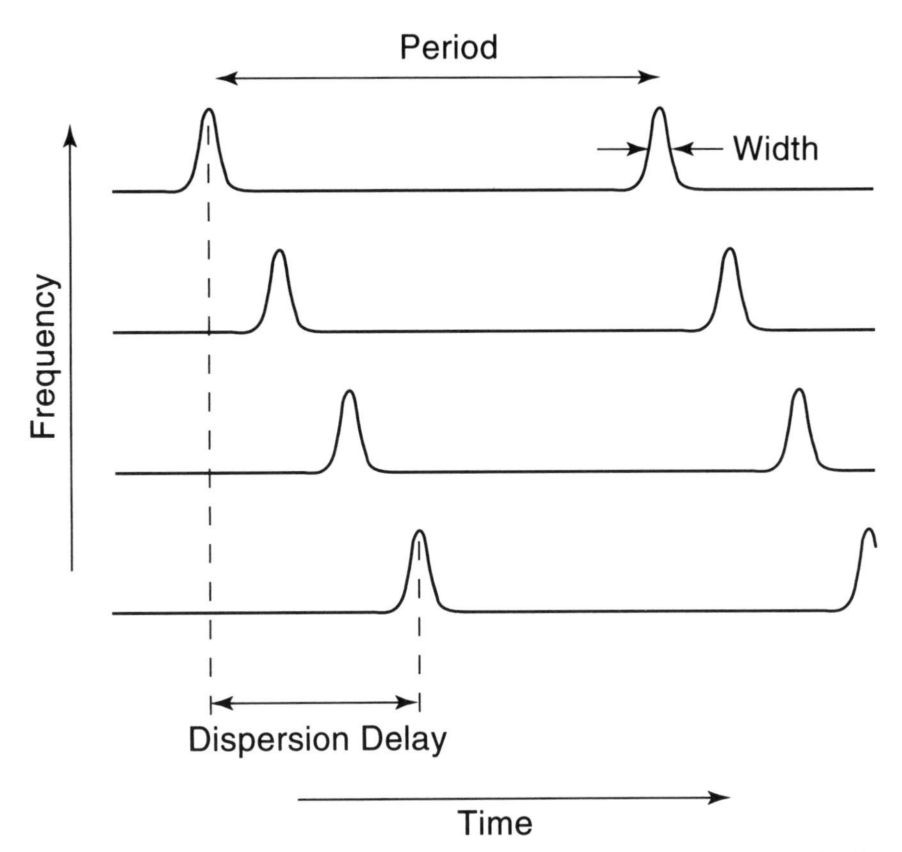

Fig. 3: A depiction of a received pulsar signal as a function of frequency and time, showing the three parameters employed in the pulsar search analysis: dispersion, pulse period, and pulse width.

each pulsar has already been discussed, while the pulse width has been adopted as the essential parameter describing the average pulse profile shape.

Since for a search we cannot know the pulsar's dispersion, pulse period, and pulse width in advance, in order to carry out an optimized detection we need to sweep across a wide range of possible values of these three parameters. This signal analysis also has to be done at every point in the sky. Adding the two stellar coordinate position parameters to the three pulsar signal parameters yields, in an overall sense, a 5-parameter search space.

Thus, the essence of obtaining the maximum possible sensitivity out of the new pulsar search was the ability to efficiently carry out an intensive and comprehensive search through the raw telescope data over this 3-dimensional parameter space of dispersion, period and pulse width. In turn, this meant that the essential new feature of this pulsar search relative to what had been done before lay in the computer analysis, not the telescope or other hardware.

DE-DISPERSION

Radio signals from pulsars suffer dispersion during their passage to the earth through the free electron density in interstellar space. This effect causes the pulses to arrive first at higher radio frequencies, and then successively later at lower frequencies, as illustrated in Figure 3. The magnitude of this effect is proportional to the integrated free electron density along the path between the pulsar and the earth. This column density is referred to as the dispersion measure (DM), given in units of $cm^{-3}pc$.

Dispersion produces a significant special problem for pulsar observations. We always wish to use a large receiver bandwidth when observing a radio source so as to improve the sensitivity of the observations. The magnitude of this dispersion effect, at the frequencies commonly used for pulsar observations, is such that the arrival time of a pulse is significantly delayed across even a typical bandwidth of a few megahertz. Hence without special processing, the pulse would be smeared out in time when detected by the receiver. For example, when observing at 430MHz at Arecibo, the differential time delay across the available 8MHz bandwidth for a distant pulsar with DM $\sim$ 200 $cm^{-3}pc$ would be 170 ms. Since this dispersion across the receiver bandwidth is much larger than a typical pulsar pulse width or even the entire period of short-period pulsars, "de-dispersion" is usually part of any pulsar observation. This is accomplished by first using a multichannel receiver to split the wide total observed bandwidth into narrower individual adjacent channels such that the dispersion smearing within each channel is now acceptable for the pulsar study in question. Successively longer time delays are then added to the data streams from each higher frequency channel in order to restore the proper time relationship between the pulses before summing all of the channels together.

While de-dispersion is a relatively straightforward process when observing a known pulsar at a given dispersion measure, a pulsar search must generate many de-dispersed data streams simultaneously, covering all possible dispersion measures of interest. This requires that the incoming data streams from the multichannel receiver must be added together not just with one set of delays added to successive channels, but with a whole range of sets of delays. An algorithm called TREE (by analogy with its computational structure) was used to carry out the required delays and summations on the incoming data as efficiently as possible by eliminating redundancies between the calculations required for different sets of delays.

This search de-dispersion processing is seen as the first step in the pulsar search data flow diagram shown in Figure 4. With the Arecibo telescope observing the sky at 430MHz, the multichannel receiver divided the 8MHz observing bandwidth into 32 adjacent 250kHz bandwidths. Detected signals from each of these channels were then sampled, de-dispersed, and written to magnetic tape in real time by the program ZBTREE running on the Modcomp II/25 minicomputer as the telescope scanned the sky.

It is interesting to note that a TREE algorithm can be used for a pulsar search all by itself, by simply scanning the resulting data streams for the characteristically dispersed pulses from a pulsar. An excess of strong pulses or an increase in the fluctuating signal power produced by weaker pulses occurring at some non-zero dispersion measure serves as the detection criterion. Such an approach, which relies only on the dispersed property of a pulsar signal, is much less sensitive than the search described here which includes a full periodicity analysis. But, in fact, it was just a search of this much simpler kind which was originally contemplated for Arecibo, and for which Joe Taylor had focused his funding proposal to the National Science Foundation for support to purchase a new computer. Once this new minicomputer arrived, however, it provided sufficient processing power to make a full de-dispersion and periodicity search a tantalizing possibility. The decision to attempt this much more ambitious goal was a critical factor in the ultimate success of the Arecibo search.

A vestige of the original dispersed pulse search was in fact retained in the REPACK program. This intermediate processing step on the Arecibo Observatory's CDC 3300 computer was needed since the Modcomp had no disk drive, and hence did not have enough space to re-organize ZBTREE data into the 136.5 second long data blocks required for the periodicity analysis program. REPACK displayed histograms of the signal strength distributions in each of the de-dispersed data channels in order to keep open the possibility of discovering unusually erratic pulsars or other types of objects which might generate sporadic non-periodic pulses. In practice, no discoveries were made this way, although the histograms did prove useful for monitoring the de-dispersion processing and indicating when severe terrestrial interference (both man-made as well as from lightning storms) was present in the data.

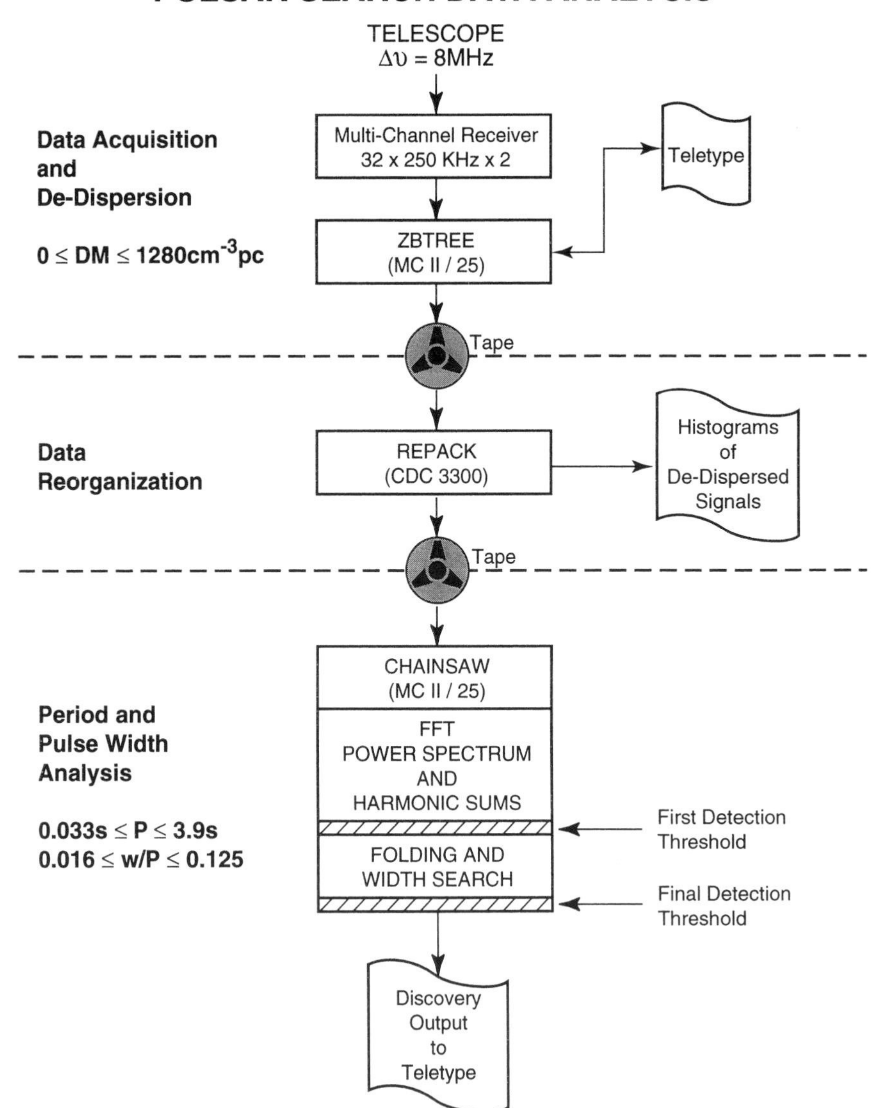

Fig. 4: A flow chart summarizing the pulsar search data analysis.

PERIOD AND PULSE WIDTH ANALYSIS

Again referring to Figure 4, we see that for the final periodicity and pulse width analysis the magnetic tapes written by REPACK were read back into the Modcomp II/25 minicomputer off-line between search observations using a program called CHAINSAW (since it reduced TREE data, of course—I think Joe Taylor gets the "credit" for coming up with this particular name).

This periodicity analysis was by far the most computationally intensive part of the overall search, and a great deal of effort was focused on finding an algorithm that would be efficient enough to meet the time constraints of the search while not sacrificing ultimate sensitivity.

The algorithm eventually developed for the period and pulse width search task was a hybrid approach, combining both Fourier analysis and folding of the data. In the Fourier analysis stage, a Fast Fourier Transform (FFT) was first used to generate a power spectrum from each data integration. This accomplished, the next task was to deal with the fact that the frequency spectrum of a periodic train of narrow pulses, such as that from a pulsar, presents the signal energy distributed across many harmonics of the fundamental pulsation frequency, with the number of harmonics present inversely proportional to the pulse width. Harmonics were thus summed within this power spectrum, with separate sums carried out over different possible numbers of harmonics for each fundamental frequency in order to cover the desired range of periods and pulse widths.

In the second phase of the analysis, the strongest signals revealed by the harmonic sums across the power spectrum were identified. This effectively formed a first detection threshold, which served to eliminate from further consideration the vast majority of the period and pulse width parameter space for each set of data which had no significant likelihood of containing a pulsar signal. The periods for each of these strongest "internal suspects" for each dispersion measure were then used to carry out a full folding and pulse width fitting analysis, with a signal-to-noise ratio calculated for each. If any of these signal-to-noise ratios exceeded a specified final threshold value, they were printed out to the teletype as possible pulsar discoveries. This final threshold was typically set at 7 σ (a 7 standard deviation signal-to-noise ratio), which may seem to be a very high value. This high threshold was necessary, however, in order to keep the false detection rate reasonably low given the large parameter space being searched.

PUTTING IT ALL TOGETHER

In the end, the computerized search system shown in Figure 4, when combined with the Arecibo telescope, achieved a pulsar detection sensitivity over ten times better than that reached by any previous pulsar search[1]. At each point in the sky scanned by the telescope, the search algorithm examined over 500,000 combinations of dispersion, period, and pulse width in the range of $0 < \mathrm{DM} < 1280\ \mathrm{cm}^{-3}\mathrm{pc}$, $.033\mathrm{s} < \mathrm{P} < 3.9\mathrm{s}$, and $.016 < \mathrm{w/P} < 0.125$.

Having described what the pulsar search system looked like on paper, let me now show you what it looked like embodied in a real computer down at Arecibo. Figure 5 is a picture of me with the Modcomp computer in the Arecibo telescope control room. I will confess that looking at this picture does make me feel a little old. This reaction comes not so much when I look at myself in the picture, but rather when I look at this computer, which I still remember thinking of as an impressively powerful machine. While it was certainly quite powerful as used for this purpose at that time, computer technology has certainly come a long way since I was working down at Arecibo.

You may note that the computer is housed in two rather crude looking wooden boxes. I made these up out of plywood at the University of Massachusetts to form a combination packing crate and equipment cabinet for shipping the computer to Arecibo, an example of the range of skills that graduate students often employ in doing their thesis work. You may also be able to see hash marks on the side of the crate near the top, which recorded a running tally of the pulsars discovered by the system.

The computer itself had a core memory of 16K 16 bit words. There was no floating point hardware, as all the calculations were done in scaled integer representations for speed. The teletype served for interactive input/output, while the tape drive provided the necessary data mass storage and the ability to communicate with other computers.

What you can't see in the picture is that this machine was programmed entirely in assembly language. Over 4000 statements on punch cards were eventually written for the analysis codes. Also, due to the limited hardware on the system (16K memory, no disk drive), I could not use the available operating systems provided by the manufacturer, which would have been too large and slow for the purposes of this application anyway. Hence, the task of programming the computer also involved writing my own custom drivers for the peripherals, servicing the interrupts, and the like. I do recall

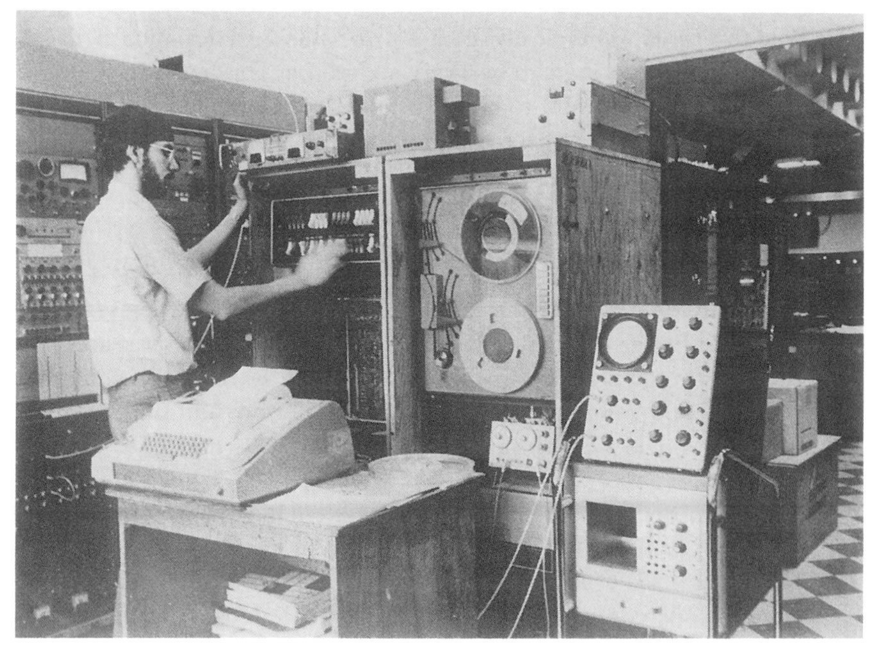

Fig. 5: A picture of myself with the Modcomp II/25 minicomputer used for the pulsar search at Arecibo, taken in the Arecibo control room shortly after the discovery of the binary pulsar.

a certain sense of pride that every bit set and every action carried out by that machine was explicitly controlled by code that I had written. I also recall saying when it was done that while it had been quite a valuable and interesting experience to understand and program a computer so extensively at such a fundamental level, once in a lifetime for this type of experience was enough !

SCANNING THE SKIES WITH COMPUTER AND TELESCOPE

As it happened, the Arecibo telescope was undergoing a major upgrade at the time we were proposing to use it for our search. Fortunately, while many other types of observations were made impossible during the construction work that this entailed, pulsar searching was not severely limited by the degraded telescope capabilities during this time. Also, as observing time had to be fit in between construction activities, a graduate student who could spend long periods living on site could take advantage of any available time on an *ad hoc* basis. Hence we had the opportunity to spend many more hours pulsar searching than I think we ever expected at the beginning of the search. At one point, I remember someone asking how much longer I would be staying down at Arecibo, to which Joe quipped "he's been down there for just one more month for the last year". In fact, I ended up at the observatory carrying out the pulsar search intermittently from December 1973 to January 1975, with time at Arecibo interspersed with periodic trips back to Massachusetts when observing time was not available.

Of course, my long days at Arecibo yielded a wide range of experiences, some of which are of the sort which are funny now but which weren't so funny then—such as the various occasions on which equipment failed, interference ruined entire observing sessions, and the like.

Coping with interference is a central fact of life for a pulsar astronomer since the relatively low radio frequencies used for these observations are plagued with interference of every type. For example, I can recall trying to eliminate one persistent long period "pulsar", which was ultimately traced to an arcing aircraft warning light on one the towers which support the telescope platform structure. Lightning storms often led to the wholesale loss of entire observing runs, and when the U.S. Navy decided to hold exercises off of the coast, there was no point even trying to take data—I just sat in the control room watching signals from the naval radars (or whatever) jump around on the observatory spectrum analyzer.

As I alluded to before, successfully carrying out one's thesis often requires one to employ skills which weren't quite part of the original job description. In my case, this included the necessity of becoming an amateur computer field service engineer, when minicomputer components including the power supply, tape drive controller, and analog-to-digital converter multiplexer failed at different inopportune moments.

Just using the Arecibo telescope was quite a memorable personal experience in and of itself. I particularly liked the fact that I could set up the telescope controls myself when doing a run, watching out the picture window in the control room as the huge feed structure moved in response to my commands to start an observation. It contributed to a feeling of direct involvement in the observations, of gazing into the heavens with some immense extension of my own eyes. At some telescopes, the control room is situated such that you can't even see the telescope during the observations. While this is hardly a practical problem for the observations, it is somewhat more romantic (and also reassuring to an astronomer's confidence that all is well) to be able to see the telescope move as you observe with it !

In the end, the pulsar search system performed beautifully. At the conclusion of the search, 40 new pulsars[2] had been discovered in the roughly 140 square degrees of sky covered in the primary search region near 19^h right ascension where a part of the plane of our galaxy is observable with the Arecibo telescope. These 40 were accompanied by the detection of 10 previously known pulsars, giving the search a 4-to-1 success rate in multiplying the pulsars known in this region. While for me this would have been quite a satisfying achievement and formed the basis for a nice thesis, it was of course eclipsed by the discovery of what was to become by far the most remarkable of these 40 new pulsars, PSR 1913+16.

MAKING DATA MAKE SENSE: THE BINARY PULSAR

By July of 1974 the pulsar search had settled down to something of a routine. I had even by then typed up and generated blank copies of various "forms" to be filled out to catalog and organize the discovery, confirmation, and subsequent period improvement observations for each new pulsar. Figure 6 shows the teletype output for the discovery observation of one such new pulsar, from data taken July 2, 1974. This output from the search system presented a succinct summary of vital information, including the pulsar position. The coordinates of this particular discovery correspond to those of PSR 1913+16, which we now also know as the binary pulsar. In this case two discovery outputs were generated, showing a signal with the same periodicity appearing in both dispersion channels 8 and 10. The pulsar period is shown on the output as about 53 ms, but my handwritten notation added to the output at the time corrects this to the true period of 59 ms. This standard correction factor of 1.1112 for the search output

Fig. 6: The teletype output from the off-line analysis program showing the discovery of PSR 1913+16 using data taken on July 2, 1974. Note that due to a shift of the sampling rate of the search to avoid 60Hz power line interference, the 53 ms pulsar period typed by the analysis program needed to be adjusted to arrive at the correct 59 ms value I noted at the time by hand on the printout.

reflected a shift of the system sampling rate from 15ms to 16.668 ms in order to avoid interference generated by the 60Hz power line frequency.

The number 7.25 on the right of the output line is of particular note, as it shows that the initial detection of the binary pulsar occurred at only 7.25 σ. Since the final search detection threshold was set at 7.0 σ for this as most other observations, had the pulsar signal been only slightly weaker on this day it would never have even been output by the computer. The initial discovery of the binary pulsar was a very close call indeed ! In retrospect, all that work focused on getting every last bit of possible sensitivity out of the pulsar search algorithms had indeed been worth the effort.

The discovery form from my notebook for PSR 1913+16 is shown in Figure 7. It shows the original 7.25 σ detection on scan 473, along with subsequent confirming re-observations on scans 526 and 535 which resulted in much more comfortable signal-to-noise ratios. The "fantastic" comment at the bottom refers to the 59 ms period of this pulsar, which at the time made it the second fastest pulsar known, the only faster one being the famous 33 ms pulsar in the Crab Nebula. At that point in time, I had no idea what lay ahead for this pulsar. However, a clue as to forthcoming events is evident on this form. You will note that as later observations of this pulsar proceeded, I kept going back to this discovery form changing the entry for the period of this pulsar, eventually crossing them all out in frustration.

After a group of new pulsar discoveries had been accumulated by the search, an observing session would be devoted to obtaining more accurate values for the pulsar periods. The first attempt to obtain a more accurate period for PSR 1913+16 occurred on August 25, 1974, nearly two months

PSR 1913+16B*

RA	DEC	PERIOD	DM
~~19:13:44~~	16°00'	.059 0332	~~150~~
19 13 13	16°00'24"	~~10698~~	167±5
±4ˢ	±60"	~~58 9824~~	
		.059030	

SCAN	DUMPS	DM CHNL	FREQ/tree	S/N
473	705-736	8,10	430/2B	7.25
526	1-128	8,9	430/2B	13.0
535	1-384	8	430/2B	15.0

ℓ = 49.95 b = 2.11 fantastic!

Fig. 7: The "discovery form" from my notebook recording the detection and confirmation of PSR 1913+16.

after the initial discovery. The standard procedure involved making two separate 5 minute to 15 minute observations for each pulsar, one near the beginning and the other near the end of the approximately 2 hour observing window that Arecibo provided for a given object each day. The data was then folded separately to arrive at a period and absolute pulse arrival time within each of these two independent observations. After correcting for the effect of the Doppler shift due to the earth's motion, the periods determined within these two integrations were accurate enough to then connect pulse phase between the observations using the two measured arrival times, providing an accurate period determination across this one or two hour baseline.

As you can see in Figure 8, my first attempt to carry out this by then standard procedure for PSR 1913+16 on August 25 produced a completely perplexing result. I always routinely compared the periods found within each of the two different short observations before proceeding with the remainder of the calculations, just to double check that nothing was amiss. As shown in my notes, something was indeed amiss for this data—and quite seriously so! Instead of the two Doppler-corrected periods being the same to within some small expected experimental error, they differed by 27 microseconds, an enormous amount. My reaction, of course, was not "Eu-

PERIOD REFINEMENT FOR PSR 1913+16B

DATE 8/25/74

RA 19 13 14 DEC 16 00'

CDC INTERFACE SAMPLES EVERY 5 MSEC FROM 10 MSEC TREE, CHNL 14

SCAN	AST START	# RECORDS	2209 PERIOD	2209 ZEROLAG	DOPPLER
567	20 46 30	180	.059 028 056	.0176	.999 950 152
574	22 11 10	180	.059 000 696	.0539	.999 949 530

FWHM ~ 15 msec (badly smeared - 2209 has ~~period~~ problems)

check agreement between 567 and 574. $\Delta p \sim \frac{(.015)(.06)}{900} \sim .000001$

$P_{\text{BC }567} = .059\ 025\ 113\ 5$

$P_{\text{BC }574} = .058\ 997\ 718\ 2$

error .000 027 — so obviously 2209 is not able to get a good period, due to sine wave nature of pulse

$\Delta n \sim (.000\ 027)\frac{8516}{.059} \sim 2.4$ so extrapolation not possible

Fig. 8: A "period refinement form" from my notebook, showing the failure of my first attempt to obtain an improved period for PSR 1913+16.

reka—its a discovery" but instead a rather annoyed "Nuts—what's wrong now?" After a second attempt to carry out the same observation two days later resulted in even worse disagreement, I determined that I was going to get to the bottom of this problem, whatever it was, and finally get a good period for this one recalcitrant pulsar.

While I really could not imagine what specific instrumental or analysis problem could be producing such an error, there was the fact that at 59 ms, PSR 1913+16 was by far the fastest pulsar I had observed. Its pulses were thus not very well resolved by the 10 ms data sampling I was using for these observations, so I speculated that perhaps this had something to do with the problem. The difficulty with going to higher time resolution was that I was using the ZBTREE search de-dispersion algorithm as part of the period refinement observations, and the search computer simply could not execute ZBTREE any faster than the 10 ms rate I was already using. So getting higher time resolution meant setting up a special observation for this one pulsar, and writing a special computer program for the Arecibo CDC 3300 computer to de-disperse the new data and format it for analysis.

Data was first taken using the new observing system on September 1 and 2, about a week after the problems with the period determination became evident. As I worked on the new system, I tried to convince myself that as soon as the new, higher time resolution observations were available, the problems would just go away. The new data, of course, yielded just the opposite result—the problem was now even worse, since the period was still changing, but now poor time resolution could no longer be blamed. Figure 9 shows a plot taken from my notebook of this first period data taken with the new system. PSR 1913+16 was observed for as long as possible on each of these two days, and the observed period of the pulsar is shown versus time for the roughly 2 hours the pulsar was observable each day. I had color-coded the data from the two days in my notebook, which does not

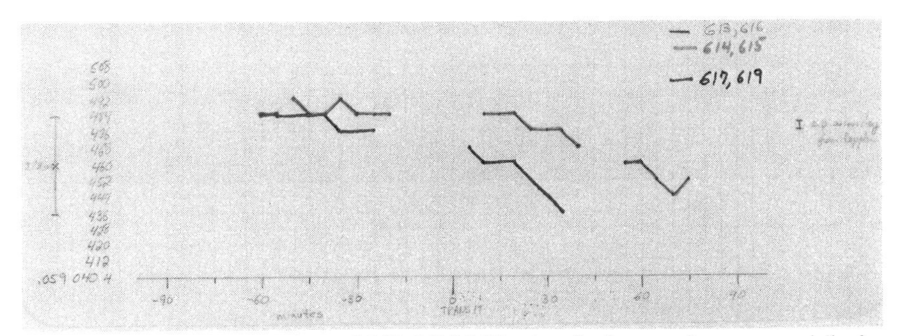

Fig. 9: The first two observations of PSR 1913+16 using the improved system specially designed to resolve the difficulties with determining the pulsation period for this pulsar. The observed pulsation period in successive 5-minute integrations is plotted versus time before and after transit. A calculation showing the magnitude of the change in the earth's Doppler shift is also seen on the right. Looking at this plot of data from September 1 and September 2, I realized that by shifting the second of these curves by 45 minutes the two curves would overlap. This was a key moment in deciphering the binary nature of PSR 1913+16.

reproduce well here in the figure. The data from the first day forms the broken upper curve, while data from the second day forms the uniformly lower broken second curve. The breaks in the curves were due to unavoidable gaps in the data. Both curves show the same behavior: the period starts out high, and then drifts to progressively lower values. The magnitude of the change of the earth's Doppler shift during this time is indicated by the small error bar on the right of the curve; so even if this correction had been done completely incorrectly, it could not really be a factor in what I was seeing here. Such a consistent drift during the course of an observation or experiment is usually quite suspiciously indicative of an instrumental problem, but, by this point, I had no good ideas left as to what could produce this effect. Certainly the observatory's precision time standards could not be suspected of drifting this much! But then, looking at this plot, it struck me that the two curves, while exhibiting the same overall downward trend, were really not identical, but that they would indeed match if shifted 45 minutes with respect to each other. While I clearly recall this as a crucial insight into the reality of the period variations, I do not recall whether or not the entire picture of PSR 1913+16 being in a binary system was also immediately clear to me at the same instant. But certainly within a short period of time I was sure that the period variations that I had been seeing were in fact due to Doppler shifts of the pulsar period produced by its orbital velocity around a companion star.

I immediately focused my observing schedule on getting as much more data on PSR 1913+16 as I possibly could, as fast as I possibly could. But a certain caution still remained; this was quite a remarkable assertion to be making, and I wanted to be absolutely confident of my conclusions before I announced my results. I thus set myself a strict criterion. If the pulsar was indeed in a binary, at some point the period would have to stop its consistent downward trend during each day's observations, reach a minimum, and start increasing again. I decided to wait for this prediction to come true to assure myself beyond all doubt that I correctly understood what I was seeing.

I had about two weeks to wait before I was able to obtain the final bit of evidence that I needed. The next page in my notebook shown in Figure 10 shows the new data as it was accumulated over this time. As before, the observed pulsar period is plotted versus time during the observing run, with data from the first two days of Sept. 1 and Sept. 2 seen copied over near the top of the page. At the lower right is the curve from Sept. 16, finally showing what I had hoped to see, the period reaching a clear minimum and then reversing direction.

By Sept. 18, my analysis of the data taken Sept. 16 showing the period minimum was complete, and I wrote a letter to Joe back in Amherst to tell him the amazing news that the pulsar was in a high-velocity binary orbit with about an 8 hour period. (Even at that point I referred to him as "Joe", rather than "Professor Taylor"—an informality that marked his approach to his students as colleagues rather than subordinates which I greatly

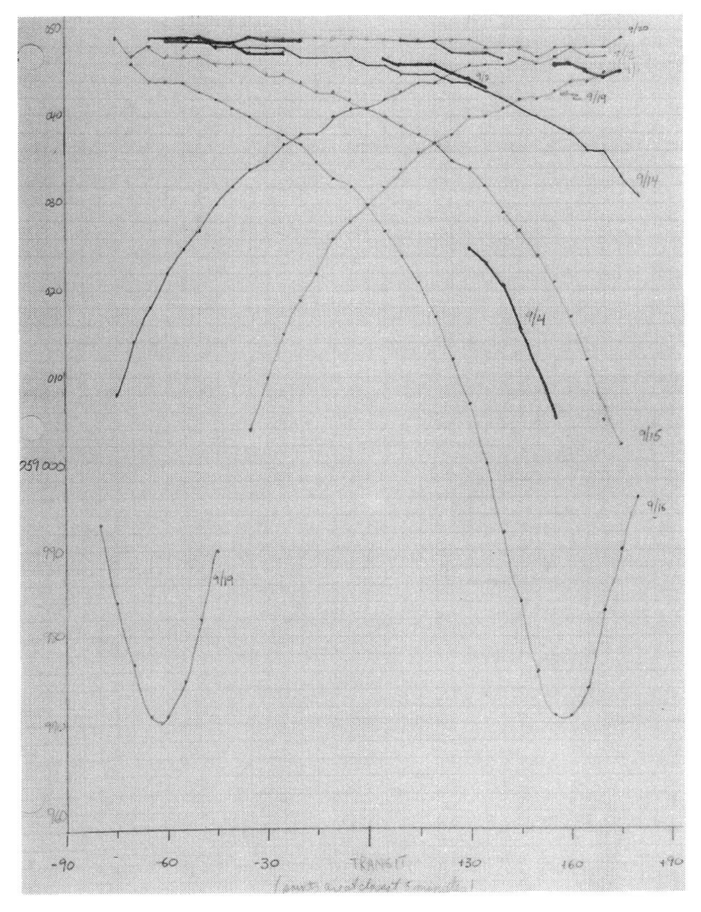

Fig. 10: After the realization that PSR 1913+16 was probably in a binary system, data was taken as often as possible on succeeding days to fully confirm this hypothesis. The criterion that I established for myself was that I would need to see the period derivative change sign and start to increase. This final confirmation of the binary hypothesis was obtained with the September 16 data shown at the lower right.

appreciated.) But just sending a letter seemed rather inadequate for such news, and as telephone connections from Arecibo were somewhat problematical, I decided to try calling him using the observatory's shortwave radio link to Cornell University. This radio transceiver communicated with its counterpart in Ithaca, New York, where a secretary could then patch the radio over into the telephone and connect me to Joe at the University of Massachusetts in Amherst. I do not recall exactly what we said, beyond telling him the news, but as you can imagine Joe was on a plane to Arecibo in very short order. He arrived at Arecibo with a hardware de-disperser which allowed the continuing observations of the binary to proceed much more efficiently than had been possible using the *ad hoc* system I had put together for the first observations.

Upon his arrival, we compared our thoughts on the status and potential of the binary pulsar discovery and set up the new equipment. Soon Joe started to program a more formal least squares fitting procedure for the orbital analysis, while I concentrated on accumulating and reducing more period data from the pulsar and finishing up the pulsar search work needed for my thesis. After Joe left to head back to Amherst, the new binary pulsar data was transferred between the pulsar data acquisition system (me) and the orbit analysis system (him) by my reading long series of numbers to him over the shortwave radio, a rather low-bandwidth but straightforward data link.

ANALYZING THE ORBIT WITH NEWTON AND EINSTEIN

The binary pulsar system revealed by the period variations which I have just shown is depicted in Figure 11. Instead of existing as an isolated star as was the case for all previously known pulsars, PSR 1913+16 is in a close elliptical orbit around an unseen companion star. In making this sketch, I have allowed myself the benefit of an important additional piece of information which was not immediately known at the time of the discovery. As Joe Taylor will describe, the pulsar and its companion are now known to have almost identical masses, and hence their two orbits are drawn as being closely comparable in size.

Combining the observed period variations shown in Figure 10 with subsequent other observations provided the first full velocity curve for the binary pulsar orbit. These measurements of the pulsar period converted to their equivalent radial velocities are shown as the data points in Figure 12, reproduced from the binary pulsar discovery paper[3] published in January 1975. These velocities are shown plotted versus orbital phase, based on the 7^h45^m period of the pulsar's orbit.

That the binary's orbital period is almost exactly commensurate with 24 hours explains why the observed period variations had shown their tantaliz-

PPPL#93X0345B

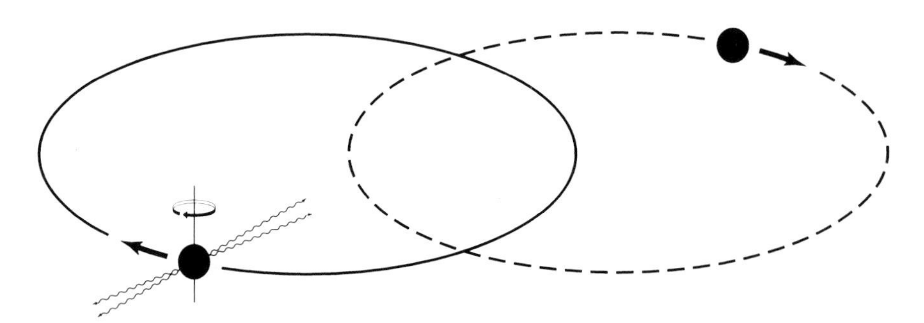

Fig. 11: A sketch of the pulsar in its elliptical orbit around a similarly massive companion star.

ing 45 minute shift each day. By pure chance the binary's orbital period was such that each time the pulsar was observable overhead on each successive day at Arecibo, it had also come around again to almost exactly the same place in its orbit. The result was that I would see almost exactly the same period variation all over again during the next day's observation. The critical 45 minute shift identified in Figure 9 thus was just the difference between three whole 7^h45^m orbits and 24 hours. Had Arecibo been a fully steerable telescope, such that I could have observed the pulsar for a longer time each a day, there would have been much less mystery and drama involved in the identification of the pulsar as a binary. But at Arecibo I did have two critical advantages which offset this problem. In the first place, the telescope's enormous sensitivity allowed me to first discover and then re-observe the pulsar in relatively short integrations, short enough so the pulsar could still be detected despite its varying period. A more subtle advantage was that at Arecibo I had the rare opportunity to use a "big science" instrument in a hands-on "small science" fashion. I had extensive access to telescope time, and I was able to quickly set up, repeat, and change my observations as I saw fit in pursuing the binary pulsar mystery.

Also shown in Figure 12 from the discovery paper is a curve fitted to the velocity data points using a Keplerian orbit, e.g., an orbit based on Newtonian physics. The orbital analysis used to arrive at this first formal fit to the

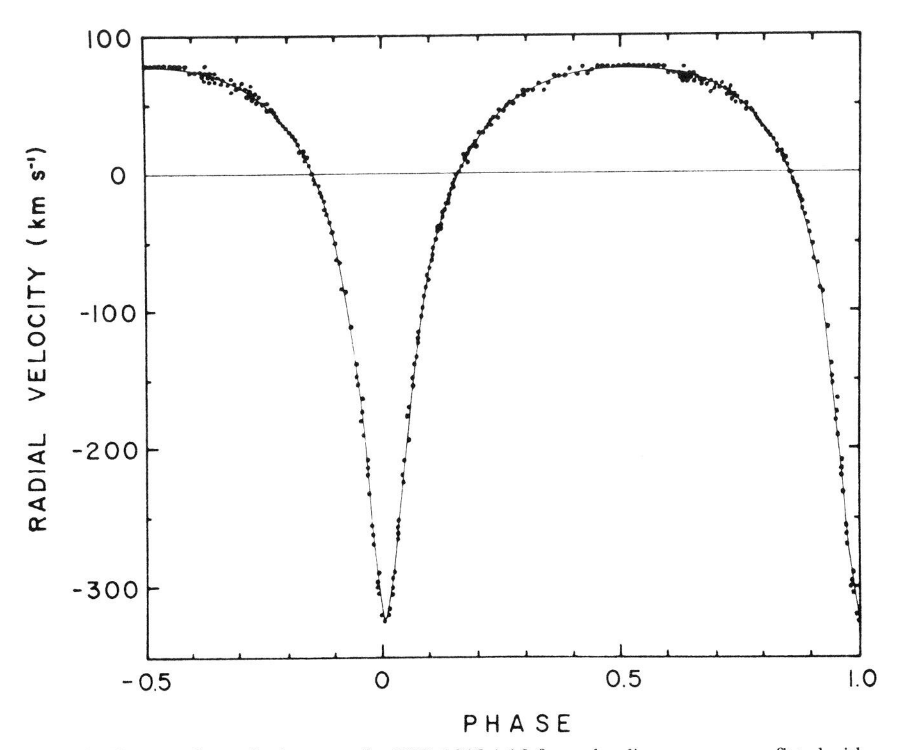

Fig. 12: The complete velocity curve for PSR 1913+16 from the discovery paper, fitted with a Keplerian orbital solution. The orbital phase is the fraction of a binary orbital period of 7^h45^m (from R.A. Hulse and J.H. Taylor, *Astrophys. J.*, 195: L51 – L53, 1975.

binary pulsar data was exactly the same as that historically applied by optical astronomers to the study of systems known as single line spectroscopic binaries. The only difference in the case of the binary pulsar was that instead of observing a periodic shift in wavelength of a line in the visible spectrum of one member of a binary star system, we were using the periodic shift of the spectral line formed by the pulsation period of the pulsar clock. Indeed, the first rough solution obtained for the orbital elements from this velocity curve involved the use of a graphical technique found in Aitken's classic book "The Binary Stars"[4], first published in 1935! This hand analysis of the data was carried out at Arecibo using a velocity curve plotted on overlapping pieces of transparent graph paper held together with paper clips. The results from this crude fit were later used as the first approximation to the orbital elements needed to initialize the more accurate least squares numerical fit shown in the paper.

This initial purely Keplerian orbit analysis has now long since been superseded by a fully relativistic solution for the orbit, determined with exquisitely high accuracy using almost 20 years of observations. The orbital elements presented in the discovery paper based on this velocity curve, however, served to dramatically quantify why we knew that relativistic effects would be important in this system immediately after the binary nature of PSR 1913+16 was identified. While the Keplerian analysis did not by itself allow a full determination of the system parameters, it certainly sufficed to yield a breathtaking overall picture of the nature of this binary system: an orbital velocity on the order of $\sim$ 0.001 the velocity of light, an orbital size on the order of the radius of our sun, and masses of the members of the system on the order of a solar mass. A succinct summary of the implications of these results was given in the discovery paper:

> "This... will allow a number of interesting gravitational and relativistic phenomena to be studied. The binary configuration provides a nearly ideal relativity laboratory including an accurate clock in a high-speed, eccentric orbit and a strong gravitational field. We note, for example, that the changes in v^2/c^2 and GM/c^2r during the orbit are sufficient to cause changes in the observed period of several parts in 10^6. Therefore, both the relativistic Doppler shift and the gravitational redshift will be easily measurable. Furthermore, the general relativistic advance of periastron should amount to about 4 deg per year, which will be detectable in a short time. The measurement of these effects, not usually observable in spectroscopic binaries, would allow the orbit inclination and the individual masses to be obtained".

As we now know, even this ambitious expectation was an understatement of the relativistic studies that would ultimately be made possible by this system. But at the time, excitement over the possibility of observing these potentially large relativistic effects was tempered by concerns over whether the system would be sufficiently clean, that is, sufficiently free of other complicating effects, to allow these measurements to be made. But by early 1975, as I was starting to write my thesis, the first quantitative evidence arrived that the system might just indeed become the amazing relativistic laboratory that it has since proved to be.

This evidence came in the form of a measurement of the advance of periastron of the orbit using a more precise pulse arrival time method to analyze the binary pulsar data. This general relativistic effect involves the rotation of the elliptical orbit itself in space. Einstein's success in explaining the observed excess 43" of arc per century advance of perihelion of Mercury's orbit in our solar system by this relativistic effect constituted one of the earliest triumphs of his theory of general relativity. Initial results now showed what certainly seemed to be this same effect in the binary pulsar orbit, but now at the incredible rate of 4 degrees per year, just as predicted in the discovery paper! Thus in the 100 years it would take Mercury's orbit to rotate by a mere 0.01 degrees, the binary pulsar orbit would rotate more than 360 degrees, turning more than once completely around! This was a dramatic confirmation indeed of the role of relativistic effects in the new binary pulsar system. In fact, it should be noted that since the binary discovery almost 20 years ago, the orbit has already rotated nearly 90 degrees due to this effect, and if the binary pulsar were to be discovered today, the shape of its velocity curve would look nothing like that shown in Figure 12 from the original discovery paper.

I will end my story of the binary pulsar at this point, historically corresponding to the end of my doctoral thesis at the University of Massachusetts. In closing, first let me thank all those at the University of Massachusetts and at Arecibo who helped make my thesis and all that has followed from it possible. I would also like to observe that the long-term study and analysis of the binary pulsar was and is an exacting task that required a maximum of patience, insight, and scientific rigor. I have always felt that there was no one more suited to this study than the person with whom I have the honor of sharing this prize. I am therefore very pleased to now turn the binary pulsar story over to someone I have admired both as a person and as a scientist for over 20 years, Professor Joseph Taylor.

REFERENCES

1. R. A. Hulse and J. H. Taylor. A high sensitivity pulsar survey. *Astrophys. J. (Letters)*, 191: L59–L61, 1974.
2. R. A. Hulse and J. H. Taylor. A deep sample of new pulsars and their spatial extent in the galaxy. *Astrophys. J. (Letters)*, 201: L55–L59, 1975.
3. R. A. Hulse and J. H. Taylor. Discovery of a pulsar in a binary system. *Astrophys. J.*, 195: L51–L53, 1975.
4. Robert G. Aitken, The Binary Stars, Dover Publications, New York, 1964.

J H Taylor

JOSEPH H. TAYLOR, JR.

I was born on March 29, 1941, in Philadelphia, Pennsylvania, the second son of Joseph Hooton Taylor and Sylvia Evans Taylor. When I was seven we moved back to the family farm in Cinnaminson Township, New Jersey, then operated by my paternal grandfather. We were three children, joined later by three more, plus two Evans cousins; like the farm's peaches and tomatoes, the eight of us grew and ripened in a healthy and carefree environment on the eastern bank of the Delaware River. Among my fondest boyhood memories are collecting stone arrowheads left on that land by its much earlier inhabitants, and erecting, together with my brother Hal, numerous large, rotating, ham-radio antennas, high above the roof of the three-story Victorian farmhouse. With one such project we managed to shear off the brick chimney, flush with the roof, much to the consternation of our parents. That incident was one of many practical lessons of my youth, not all absorbed in the most timely fashion, involving ill-advised shortcuts toward some goal.

Both the Evans and Taylor families have deep Quaker roots going back to the days of William Penn and his Philadelphia experiment. My parents were living examples of frugal Quaker simplicity, twentieth-century style; their very lives taught lessons of tolerance for human diversity and the joys of helping and caring for others. Our house was large, open, and friendly. To my knowledge it has never been (nor indeed can be) locked. In our school years, Hal and I filled most of the third floor with working ham-radio transmitters and receivers. Our rigs were mostly built from a mixture of post-war surplus equipment and junk television sets. We learned by experience that when you need high voltage, the power company's 6,000-to-120-volt transformers work admirably in reverse; and that most amplifiers will oscillate, especially if you don't want them to.

I was educated mostly at Quaker institutions, in particular Moorestown Friends School and Haverford College. In school, mathematics was my first academic love. Somewhat backward high-school introductions to chemistry and physics (I failed to recognize them as such at the time) did not dampen any enthusiasm for science, they just gave me more time for sports, then a greater passion. Soccer, basketball, baseball, golf, and tennis claimed much of my energy through the Haverford years. Concurrently, though, I began discovering the delights of what science is really about. A fascinating senior honors project in physics allowed me to combine a working knowledge of radio-frequency electronics with an awakening appreciation of scientific inquiry, and to build a working radio telescope. My principal references were an old friend, *The Radio Amateur's Handbook*, and an early book on radio astronomy by Pawsey and Bracewell. This thoroughly enjoyable honors project cannot really qualify as research—everything I accomplished had been done by others, years before—but it provided excellent lessons in problem-solving of various kinds. It also delivered a valid reason for selecting something I had been hoping to find: a desirable field of physics in which to pursue graduate studies.

My academic work in the Harvard departments of Astronomy, Physics, and Applied Mathematics was the hardest I ever remember working, at least during my first year there. I suppose every beginning graduate student feels that he or she has something to prove; anyway, I certainly did. But my thesis research in radio astronomy was, once again, thoroughly enjoyable. My mentor, Alan Maxwell, knew the field and its participants well. He gave me plenty of flexibility, provided inroads and introductions when I needed them, and taught me (among many other things) the importance of clear, well-crafted writing in a scientific paper. Ron Bracewell again played an unwitting rôle; his 1965 book *The Fourier Transform and its Applications* came out just in time to give me some crucial insights necessary for analyzing the data for my thesis. It also prepared me for understanding the signal-processing techniques that later became important in my study of pulsars.

I have noticed in recent years that many budding scientists worry much more than I ever did about what the future may bring: how to get into the best university, work with the biggest names, find the best post-doctoral fellowship, and secure the ideal university position. My own psychological bent, insofar as it has influenced any professional decisions, is to pursue a path promising enjoyment along the way, without looking too far ahead. Perhaps related to my Quaker upbringing, I've always valued personal involvement in a difficult task over appeals to eminence or authority; I like the challenge of re-examining a problem from fresh perspectives. Ultimately, I believe that in important matters we are mostly self-taught, but in a way that is strongly reinforced by cooperative human relationships. I have worked in two extremely stimulating intellectual environments, first at the University of Massachusetts and more recently at Princeton. I'm fortunate to have associated with some uniquely gifted individuals who have been especially compatible co-seekers of diverse truths and pleasures: among them Dick Manchester, Russell Hulse, Peter McCulloch, Joel Weisberg, Thibault Damour, Dan Stinebring, students too numerous to name, and especially my dearly beloved wife, Marietta Bisson Taylor.

BINARY PULSARS AND RELATIVISTIC GRAVITY

Nobel Lecture, December 8, 1993

by

JOSEPH H. TAYLOR, JR.

Princeton University, Department of Physics, Princeton, NJ 08544, USA

I. SEARCH AND DISCOVERY

Work leading to the discovery of the first pulsar in a binary system began more than twenty years ago, so it seems reasonable to begin with a bit of history. Pulsars burst onto the scene (1) in February 1968, about a month after I completed my PhD at Harvard University. Having accepted an offer to remain there on a post-doctoral fellowship, I was looking for an interesting new project in radio astronomy. When *Nature* announced the discovery of a strange new rapidly pulsating radio source, I immediately drafted a proposal, together with Harvard colleagues, to observe it with the 92 m radio telescope of the National Radio Astronomy Observatory. By late spring we had detected and studied all four of the pulsars which by then had been discovered by the Cambridge group, and I began thinking about how to find further examples of these fascinating objects, which were already thought likely to be neutron stars. Pulsar signals are generally quite weak, but have some unique characteristics that suggest effective search strategies. Their otherwise noise-like signals are modulated by periodic, impulsive waveforms; as a consequence, dispersive propagation through the interstellar medium makes the narrow pulses appear to sweep rapidly downward in frequency. I devised a computer algorithm for recognizing such periodic, dispersed signals in the inevitable background noise, and in June 1968 we used it to discover the fifth known pulsar (2).

Since pulsar emissions exhibited a wide variety of new and unexpected phenomena, we observers put considerable effort into recording and studying their details and peculiarities. A pulsar model based on strongly magnetized, rapidly spinning neutron stars was soon established as consistent with most of the known facts (3). The model was strongly supported by the discovery of pulsars inside the glowing, gaseous remnants of two supernova explosions, where neutron stars should be created (4, 5), and also by an observed gradual lengthening of pulsar periods (6) and polarization measurements that clearly suggested a rotating source (7). The electrodynamical properties of a spinning, magnetized neutron star were studied theoretically (8) and shown to be plausibly capable of generating broadband radio noise detectable over interstellar distances. However, the rich diversity of the observed radio pulses suggested magnetospheric complexities far beyond those readily incorporated in theoretical models. Many of us suspected that detailed understanding of the pulsar emission mechanism might be a long time coming—and that, in any case, the details might not turn out to be fundamentally illuminating.

In September 1969, I joined the faculty at the University of Massachusetts, where a small group of us planned to build a large, cheap radio telescope especially for observing pulsars. Our telescope took several years to build, and during this time it became clear that whatever the significance of their magnetospheric physics, pulsars were interesting and potentially important to study for quite different reasons. As the collapsed remnants of supernova explosions, they could provide unique experimental data on the final stages of stellar evolution, as well as an opportunity to study the properties of nuclear matter in bulk. Moreover, many pulsars had been shown to be remarkably stable natural clocks (9), thus providing an alluring challenge to the experimenter, with consequences and applications about which we could only speculate at the time. For such reasons as these, by the summer of 1972 I was devoting a large portion of my research time to the pursuit of accurate timing measurements of known pulsars, using our new telescope in western Massachusetts, and to planning a large-scale pulsar search that would use bigger telescopes at the national facilities.

I suspect it is not unusual for an experiment's motivation to depend, at least in part, on private thoughts quite unrelated to avowed scientific goals. The challenge of a good intellectual puzzle, and the quiet satisfaction of finding a clever solution, must certainly rank highly among my own incentives and rewards. If an experiment seems difficult to do, but plausibly has interesting consequences, one feels compelled to give it a try. Pulsar searching is the perfect example: it's clear that there must be lots of pulsars out there, and once identified, they are not so very hard to observe. But finding each one for the first time is a formidable task, one that can become a sort of detective game. To play the game you invent an efficient way of gathering clues, sorting, and assessing them, hoping to discover the identities and celestial locations of all the guilty parties.

Most of the several dozen pulsars known in early 1972 were discovered by examination of strip-chart records, without benefit of further signal processing. Nevertheless, it was clear that digital computer techniques would be essential parts of more sensitive surveys. Detecting new pulsars is necessarily a multi-dimensional process; in addition to the usual variables of two spatial coordinates, one must also search thoroughly over wide ranges of period and dispersion measure. Our first pulsar survey, in 1968, sought evidence of pulsar signals by computing the discrete Fourier transforms of long sequences of intensity samples, allowing for the expected narrow pulse shapes by summing the amplitudes of a dozen or more harmonically related frequency components. I first described this basic algorithm (10) as part of a discussion of pulsar search techniques, in 1969. An efficient dispersion-compensating algorithm was conceived and implemented soon afterward (11, 12), permitting extension of the method to two dimensions. Computerized searches over period and dispersion measure, using these basic algorithms, have by now accounted for discovery of the vast majority of nearly 600 known pulsars, including forty in binary systems (13, 14).

In addition to private stimuli related to "the thrill of the chase," my outwardly expressed scientific motivation for planning an extensive pulsar survey in 1972 was a desire to double or triple the number of known pulsars. I had in mind the need for a more solid statistical basis for drawing conclusions about the total number of pulsars in the Galaxy, their spatial distribution, how they fit into the scheme of stellar evolution, and so on. I also realized (15) that it would be highly desirable "...to find even *one* example of a pulsar in a binary system, for measurement of its parameters could yield the pulsar mass, an extremely important number." Little did I suspect that just such a discovery would be made, or that it would have much greater significance that anyone had foreseen! In addition to its own importance, the binary pulsar PSR 1913+16 is now recognized as the harbinger of a new class of unusually short-period pulsars with numerous important applications.

An up-to-date map of known pulsars on the celestial sphere is shown in Figure 1. The binary pulsar PSR 1913+16 is found in a clump of objects close to the Galactic plane around longitude 50°, a part of the sky that passes directly overhead at the latitude of Puerto Rico. Forty of these pulsars, including PSR 1913+16, were discovered in the survey that Russell Hulse and I carried out with the 305 m Arecibo telescope (16–18). Figure 2 illustrates the periods and spin-down rates of known pulsars, with those in binary systems marked by larger circles around the dots. All radio pulsars slow down gradually in their own rest frames, but the slowdown rates vary

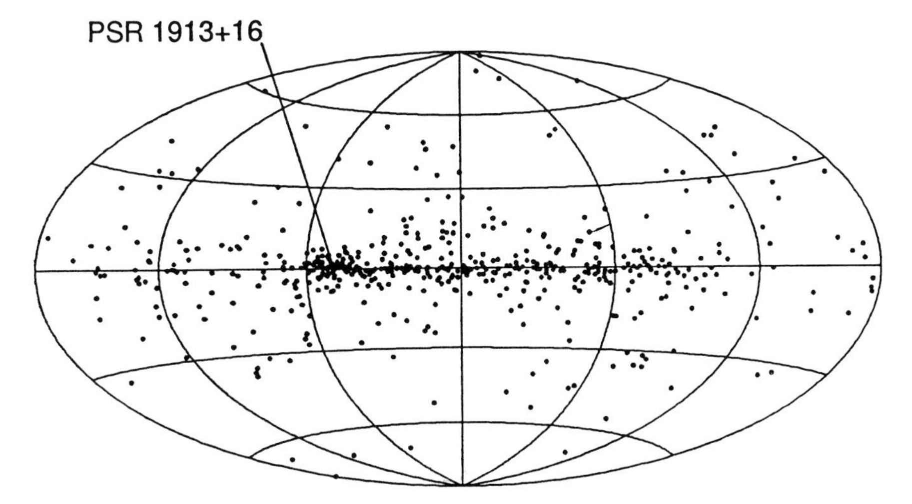

Fig. 1: Distribution of 558 pulsars in Galactic coordinates. The Galactic center is in the middle, and longitude increases to the left.

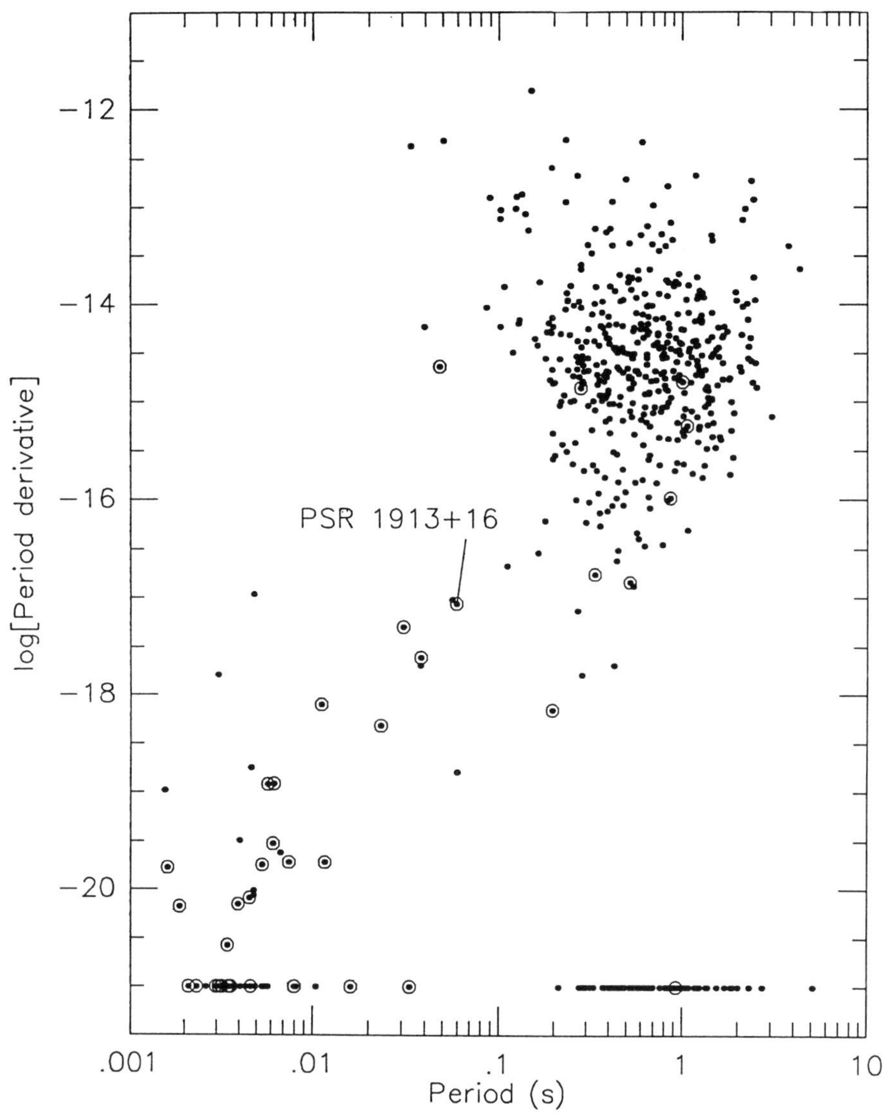

Fig. 2: Periods and period derivatives of known pulsars. Binary pulsars, denoted by larger circles around the dots, generally have short periods and small derivatives. Symbols aligned near the bottom represent pulsars for which the slow-down rate has not yet been measured.

over nine orders of magnitude. Figure 2 makes it clear that binary pulsars are special in this regard. With few exceptions, they have unusually small values of both period and period derivative – an important fact which helps to make them especially suitable for high-precision timing measurements.

Much of the detailed implementation and execution of our 1973–74 Arecibo survey was carried out by Russell Hulse. He describes that work, and particularly the discovery of PSR 1913+16, in his accompanying lecture (19). The significant consequences of our discovery have required accurate timing measurements extending over many years, and since

1974 – 76 I have pursued these with a number of other collaborators. I shall now turn to a description of these observations.

II. CLOCK-COMPARISON EXPERIMENTS

Pulsar timing experiments are straightforward in concept: one measures pulse times of arrival (TOAs) at the telescope, and compares them with time kept by a stable reference clock. A remarkable wealth of information about a pulsar's spin, location in space, and orbital motion can be obtained from such simple measurements. For binary pulsars, especially, the task of analyzing a sequence of TOAs often assumes the guise of another intricate detective game. Principal clues in this game are the recorded TOAs. The first and most difficult objective is the assignment of unambiguous pulse numbers to each TOA, despite the fact that some of the observations may be separated by months or even years from their nearest neighbors. During such inevitable gaps in the data, a pulsar may have rotated through as many as $10^7 - 10^{10}$ turns, and in order to extract the maximum information content from the data, these integers must be recovered *exactly*. Fortunately, the correct sequence of pulse numbers is easily recognized, once attained, so you can tell when the game has been "won."

A block diagram of equipment used for recent pulsar timing observations (20) at Arecibo is shown in Figure 3. Incoming radio-frequency signals from the antenna are amplified, converted to intermediate frequency, and passed

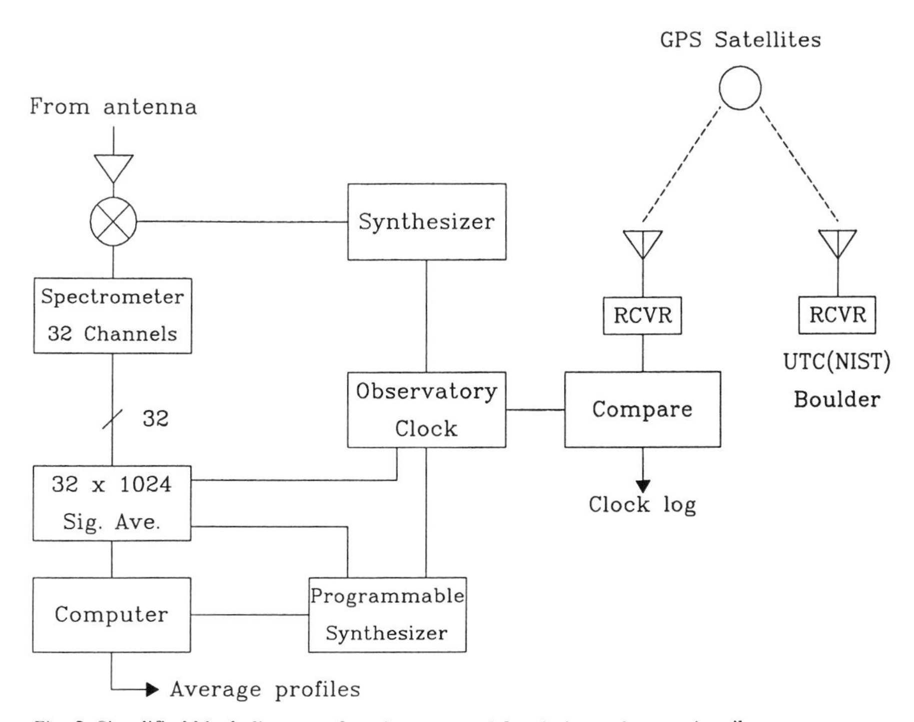

Fig. 3: Simplified block diagram of equipment used for timing pulsars at Arecibo.

through a multichannel spectrometer equipped with square-law detectors. A bank of digital signal averagers accumulates estimates of a pulsar's periodic waveform in each spectral channel, using a pre-computed digital ephemeris and circuitry synchronized with the observatory's master clock. A programmable synthesizer, its output frequency adjusted once a second in a phase-continuous manner, compensates for changing Doppler shifts caused by accelerations of the pulsar and the telescope. Average profiles are recorded once every few minutes, together with appropriate time tags. A

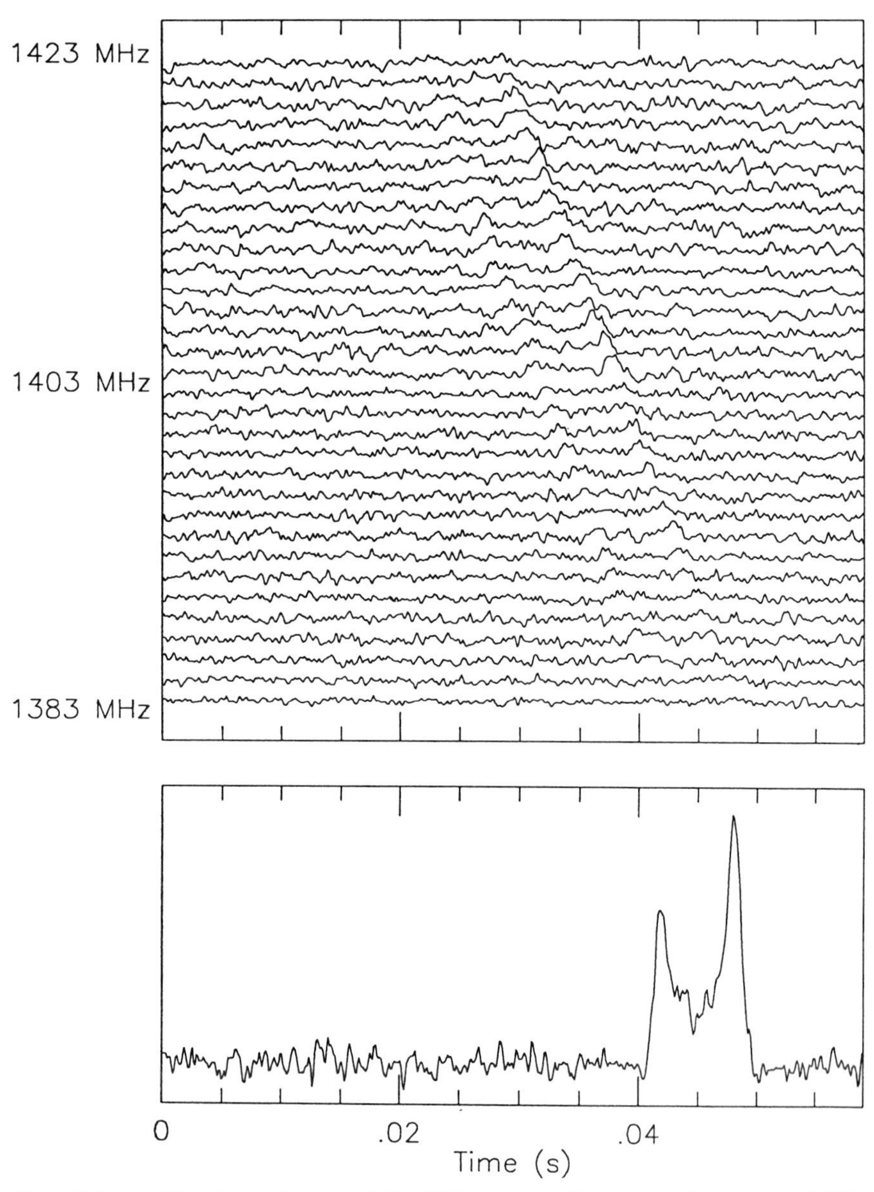

Fig. 4: Pulse profiles obtained on April 24, 1992 during a five-minute observation of PSR 1913+16. The characteristic double-peaked shape, clearly seen in the de-dispersed profile at the bottom, is also discernible in the 32 individual spectral channels.

log is kept of small measured offsets (typically of order $1\mu s$) between the observatory clock and the best available standards at national time-keeping laboratories, with time transfer accomplished via satellites in the Global Positioning System.

An example of pulse profiles recorded during timing observations of PSR 1913+16 is presented in Figure 4, which shows intensity profiles for 32 spectral channels spanning the frequency range 1383–1423 MHz, followed by a "de-dispersed" profile at the bottom. In a five-minute observation such as this, the signal-to-noise ratio is just high enough for the double-peaked pulse shape of PSR 1913+16 to be evident in the individual channels. Pulse arrival times are determined by measuring the phase offset between each observed profile and a long-term average with much higher signal-to-noise ratio. Differential dispersive delays are removed, the adjusted offsets are averaged over all channels, and the resulting mean value is added to the time-tag to obtain an equivalent TOA. Nearly 5,000 such five-minute measurements have been obtained for PSR 1913+16 since 1974, using essentially this technique. Through a number of improvements in the data-taking systems (21–26), the typical uncertainties have been reduced from around $300\mu s$ in 1974 to $15-20\mu s$ since 1981.

III. MODEL FITTING

In the process of data analysis, each measured topocentric TOA, say t_{obs}, must be transformed to a corresponding proper time of emission T in the pulsar frame. Under the assumption of a deterministic spin-down law, the rotational phase of the pulsar is given by

$$\phi(T) = \nu T + \frac{1}{2}\dot{\nu}T^2, \tag{1}$$

where ϕ is measured in cycles, $\nu \equiv 1/P$ is the rotation frequency, P the period, $\dot{\nu}$ the slowdown rate. Since a topocentric TOA is a relativistic space-time event, it must be transformed as a four-vector. The telescope's location at the time of a measurement is obtained from a numerically integrated solar-system model, together with published data on the Earth's unpredictable rotational variations. As a first step one normally transforms to the solar-system barycenter, using the weak-field, slow-motion limit of general relativity. The necessary equations include terms depending on the positions, velocities, and masses of all significant solar-system bodies. Next, one accounts for propagation effects in the interstellar medium; and finally, for the orbital motion of the pulsar itself.

With presently achievable accuracies, all significant terms in the relativistic transformation can be summarized in the single equation

$$\begin{aligned} T = t_{obs} - t_0 + \Delta_C - D/f^2 + \Delta_{R\odot}(\alpha, \delta, \mu_\alpha, \mu_\delta, \pi) + \Delta_{E\odot} - \\ \Delta_{S\odot}(\alpha, \delta) - \Delta_R(x, e, P_b, T_0, \omega, \dot{\omega}, \dot{P}_b) - \Delta_E(\gamma) - \Delta_S(r, s). \end{aligned} \tag{2}$$

Here t_0 is a nominal equivalent TOA at the solar system barycenter;

Δ_C represents measured clock offsets; D/f^2 is the dispersive delay for propagation at frequency f through the interstellar medium; $\Delta_{R\odot}$, $\Delta_{E\odot}$, and $\Delta_{S\odot}$ are propagation delays and relativistic time adjustments within the solar system; and Δ_R, Δ_E, and Δ_S are similar terms for effects within a binary pulsar's orbit. Subscripts on the various Δ's indicate the nature of the time-dependent delays, which include "Römer," "Einstein," and "Shapiro" delays in the solar system and in the pulsar orbit. The Römer terms have amplitudes comparable to the orbital periods times $v/2\pi c$, where v is the orbital velocity and c the speed of light. The Einstein terms, representing the integrated effects of gravitational redshift and time dilation, are smaller by another factor ev/c, where e is the orbital eccentricity. The Shapiro time delay is a result of reduced velocities that accompany the well-known bending of light rays propagating close to a massive object. The delay amounts to about 120 μs for one-way lines of sight grazing the Sun, and the magnitude depends logarithmically on the angular impact parameter. The corresponding delay within a binary pulsar orbit depends on the companion star's mass, the orbital phase, and the inclination i between the orbital angular momentum and the line of sight.

Figure 5 illustrates the combined orbital delay $\Delta_R + \Delta_E + \Delta_S$ for PSR 1913+16, plotted as a function of orbital phase. Despite the fact that the Einstein and Shapiro effects are orders of magnitude smaller than the Römer delay, they can still be measured separately if the precision of available TOAs is high enough. In fact, the available precision is very high

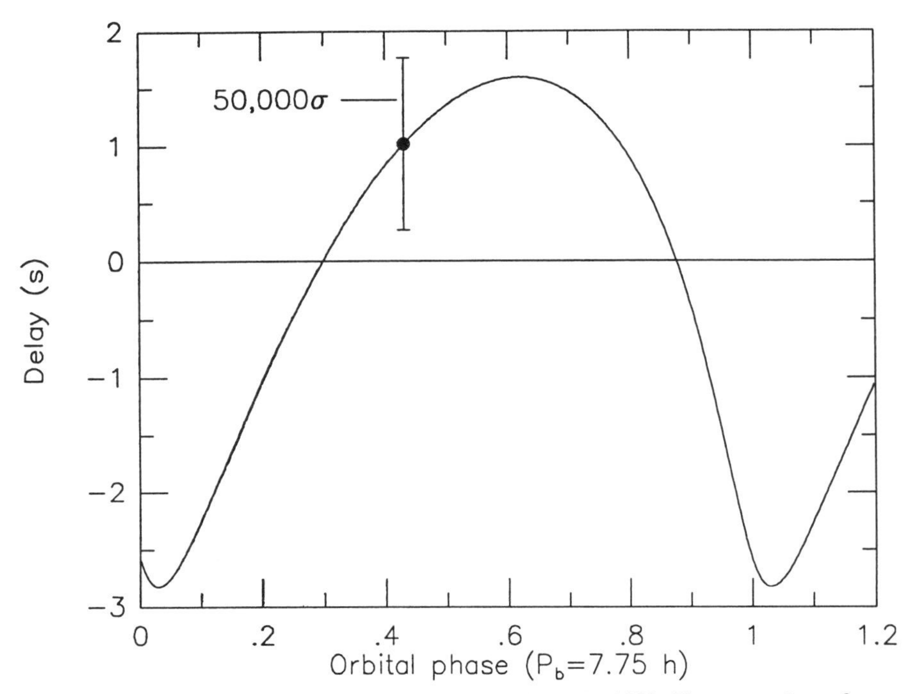

Fig. 5: Orbital delays observed for PSR 1913+16 during July, 1988. The uncertainty of an individual five-minute measurement is typically 50,000 times smaller than the error bar shown.

indeed, as one can see from the lone data point shown in Figure 5 with 50,000σ error bars.

Eqs.(1) and (2) have been written to show explicitly the most significant dependences of pulsar phase on as many as nineteen *a priori* unknowns. In addition to the rotational frequency ν and spin-down rate $\dot{\nu}$, these phenomenological parameters include a reference arrival time t_0, the dispersion constant D, celestial coordinates α and δ, proper-motion terms μ_α and μ_δ, and annual parallax π. For binary pulsars the terms on the second line of Eq.(2), with as many as ten significant orbital parameters, are also required. The additional parameters include five that would be necessary even in a purely Keplerian analysis of orbital motion: the projected semi-major axis x $\equiv a_1 \sin i/c$, eccentricity e, binary period P_b, longitude of periastron ω, and time of periastron T_0. If the experimental precision is high enough, relativistic effects can yield the values of five further "post-Keplerian" parameters: the secular derivatives $\dot{\omega}$ and $\dot{P}_b$, the Einstein parameter γ, and the range and shape of the orbital Shapiro delay, r and $s \equiv \sin i$. Several earlier versions of this formalism for treating timing measurements of binary pulsars exist (27–29), and have been historically important to our progress with the PSR 1913+16 experiment. The elegant framework outlined here was derived during 1985–86 by Damour and Deruelle (30, 31).

Model parameters are extracted from a set of TOAs by calculating the pulsar phases $\phi(T)$ from Eq. (1) and minimizing the weighted sum of squared residuals,

$$\chi^2 = \sum_{i=1}^{N} \left(\frac{\phi(T_i) - n_i}{\sigma_i/P} \right)^2, \tag{3}$$

with respect to each parameter to be determined. In this equation, n_i is the closest integer to $\phi(T_i)$, and σ_i is the estimated uncertainty of the i'th TOA. In a valid and reliable solution the value of χ^2 will be close to the number of degrees of freedom, i.e., the number of measurements N minus the number of adjustable parameters. Parameter errors so large that the closest integer to $\phi(T_i)$ may not be the correct pulse number are invariably accompanied by huge increases in χ^2; this is the reason for my earlier statement that correct pulse numbering is easily recognizable, once attained. In addition to providing a list of fitted parameter values and their estimated uncertainties, the least-squares solution produces a set of post-fit residuals, or differences between measured TOAs and those predicted by the model (see Figure 6). The post-fit residuals are carefully examined for evidence of systematic trends that might suggest experimental errors, or some inadequacy in the astrophysical model, or perhaps deep physical truths about the nature of gravity.

Necessarily, some model parameters will be easier to measure than others. When many TOAs are available, spaced over many months or years, it generally follows that at least the pulsar's celestial coordinates, spin parameters, and Keplerian orbital elements will be measurable with high precision, often as many as 6–14 significant digits. As we will see, the relativistic

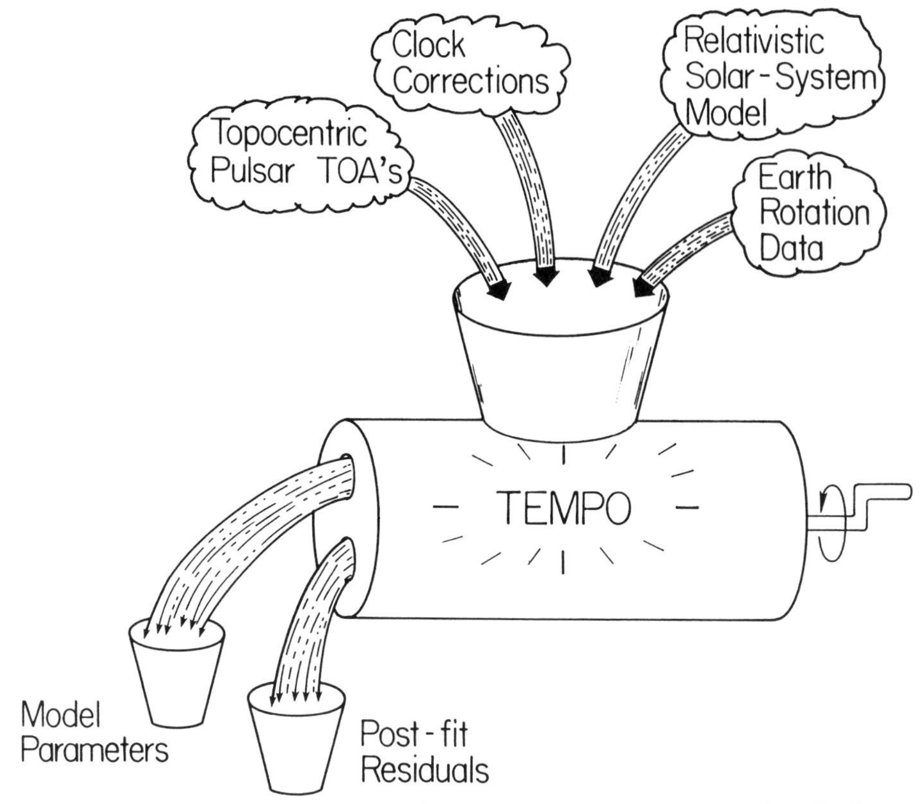

Fig. 6: Schematic diagram of the analysis of pulsar timing measurements carried out by the computer program TEMPO. The essential functions are all described in the text.

parameters of binary pulsar orbits are generally much more difficult to measure—but the potential rewards for doing so are substantial.

IV. THE NEWTONIAN LIMIT

Thirty-five binary pulsar systems have now been studied well enough to determine their basic parameters, including the Keplerian orbital elements, with good accuracy. For each system the orbital period P_b and projected semi-major axis x can be combined to give the mass function,

$$f_1(m_1, m_2, s) = \frac{(m_2\, s)^3}{(m_1 + m_2)^2} = \frac{x^3}{T_\odot (P_b/2\pi)^2}. \tag{4}$$

Here m_1 and m_2 are the masses of the pulsar and companion in units of the Sun's mass, $M_\odot$; I use the shorthand notations $s \equiv \sin i$, $T_\odot \equiv GM_\odot/c^3 = 4.925490947 \cdot 10^{-6}$s, where G is the Newtonian constant of gravity. In the absence of other information, the mass function cannot provide unique solutions for m_1, m_2, or s. Nevertheless, likely values of m_2 can be estimated by assuming a pulsar mass close to 1.4 $M_\odot$ (the Chandrasekhar limit for white dwarfs) and the median value $\cos i = 0.5$, which implies $s = 0.87$. With this approach one can distinguish three categories of binary pulsars, which I

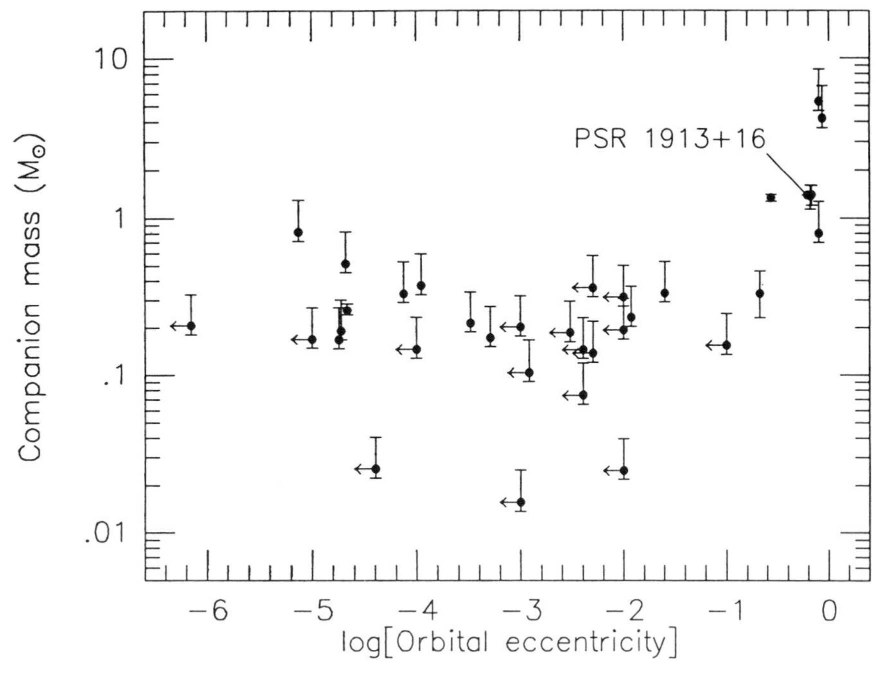

Fig. 7: Masses of the companions of binary pulsars, plotted as a function or orbital eccentricity. Near the marked location of PSR 1913+16, three distinct symbols have merged into one; these three binary systems, as well as their two nearest neighbors in the graph, are thought to be pairs of neutron stars. The two pulsars at the upper right are accompanied by high-mass main-sequence stars, while the remainder are believed to have white-dwarf companions.

shall discuss by reference to Figure 7: a plot of binary-pulsar companion masses versus orbital eccentricities.

Twenty-eight of the binary systems in Figure 7 have orbital eccentricities $e < 0.25$ and low-mass companions likely to be degenerate dwarfs. Most of these have nearly circular orbits; indeed, the only ones with eccentricities more than a few percent are located in globular clusters, and their orbits have probably been perturbed by near collisions with other stars. Five of the binaries have much larger eccentricities and likely companion masses of $0.8M_\odot$ or more; these systems are thought to be pairs of neutron stars, one of which is the detectable pulsar. Their large orbital eccentricities are almost certainly the result of rapid ejection of mass in the supernova explosion creating the second neutron star. Finally, at the upper right of Figure 7 we find two binary pulsars that move in eccentric orbits around high-mass main-sequence stars. These systems have not yet evolved to the stage of a second supernova explosion. Unlike the binary pulsars with compact companions, these two systems have orbits that could be significantly modified by complications such as tidal forces or mass loss.

V. GENERAL RELATIVITY AS A TOOL

As Russell Hulse and I suggested (17) in the discovery paper for PSR 1913+16, it should be possible to combine measurements of relativistic orbital parameters with the mass function, thereby determining masses of both stars and the orbital inclination. In the post-Keplerian (PK) framework outlined above, each measured PK parameter defines a unique curve in the (m_1, m_2) plane, valid within a specified theory of gravity. Experimental values for any two PK parameters (say $\dot{\omega}$ and γ, or perhaps r and s) establish the values of m_1, m_2, and s unambiguously. In general relativity the equations for the five most significant PK parameters are as follows (25, 31, 32):

$$\dot{\omega} = 3\left(\frac{P_b}{2\pi}\right)^{-5/3}(T_\odot M)^{2/3}(1-e^2)^{-1}, \tag{5}$$

$$\gamma = e\left(\frac{P_b}{2\pi}\right)^{1/3} T_\odot^{2/3} M^{-4/3} m_2 (m_1 + 2m_2), \tag{6}$$

$$\dot{P}_b = -\frac{192\pi}{5}\left(\frac{P_b}{2\pi}\right)^{-5/3}\left(1+\frac{73}{24}e^2+\frac{37}{96}e^4\right)(1-e^2)^{-7/2} T_\odot^{5/3} m_1 m_2 M^{-1/3}, \tag{7}$$

$$r = T_\odot m_2, \tag{8}$$

$$s = x\left(\frac{P_b}{2\pi}\right)^{-2/3} T_\odot^{-1/3} M^{2/3} m_2^{-1}. \tag{9}$$

Again the masses m_1, m_2, and $M \equiv m_1 + m_2$ are expressed in solar units. I emphasize that the left-hand sides of Eqs. (5–9) represent directly measurable quantities, at least in principle. Any two such measurements, together with the well-determined values of e and P_b, will yield solutions for m_1 and m_2 as well as explicit predictions for the remaining PK parameters.

The binary systems most likely to yield measurable PK parameters are those with large masses and high eccentricities and which are astrophysically "clean," so that their orbits are overwhelmingly dominated by the gravitational interactions between two compact masses. The five pulsars clustered near PSR 1913+16 in Figure 7 would seem to be especially good candidates, and this has been borne out in practice. In the most favorable circumstances, even binary pulsars with low-mass companions and nearly circular orbits can yield significant post-Keplerian measurements. The best present example is PSR 1855+09: its orbital plane is nearly parallel to the line of sight, greatly magnifying the orbital Shapiro delay. The relevant measurements (33–35) are illustrated in Figure 8, together with the fitted function $\Delta_S(r, s)$, in this case closely approximated by

$$\Delta_S = -2r\log(1 - s\cos[2\pi(\phi-\phi_0]) \tag{10}$$

where ϕ is the orbital phase in cycles and $\phi_0 = 0.4823$ the phase of superior conjunction. The fitted values of r and s yield the masses $m_1 = 1.50^{+0.26}_{-0.14}$, m_2

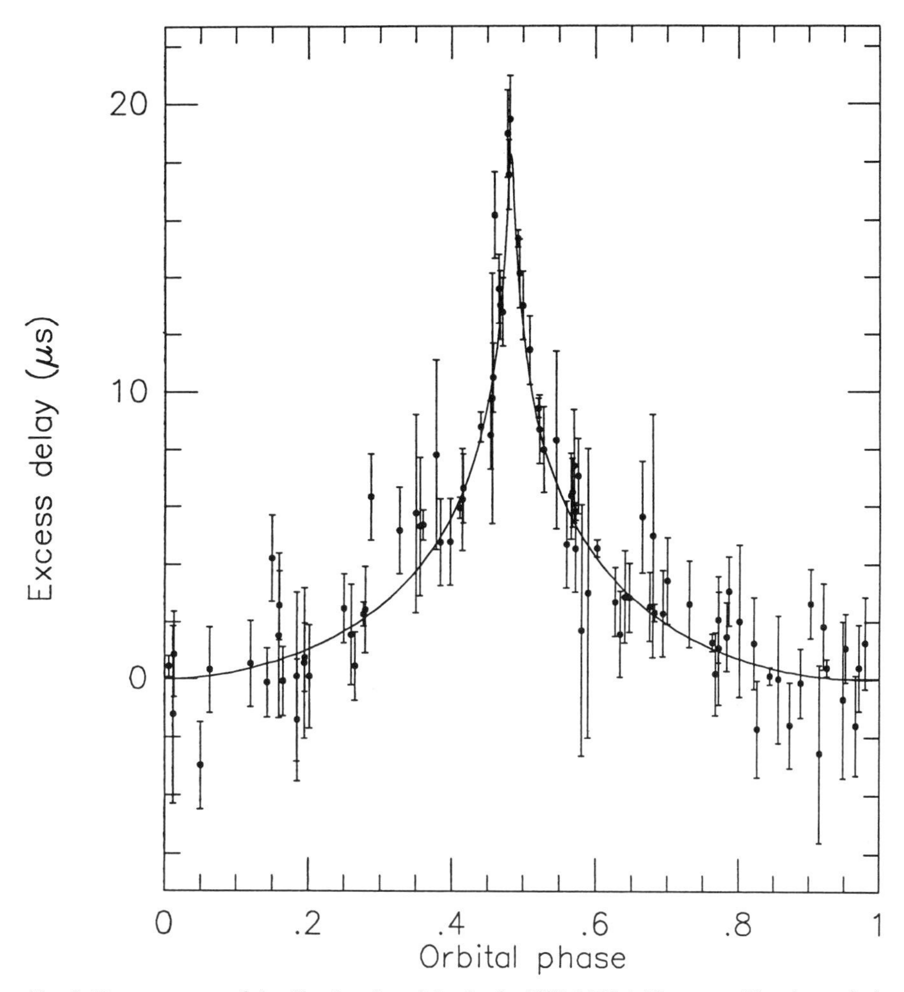

Fig. 8: Measurements of the Shapiro time delay in the PSR 1855+09 system. The theoretical curve corresponds to Eq. (10), and the fitted values of r and s can be used to determine the masses of the pulsar and companion star.

$= 0.258^{+0.028}_{-0.016}$. In a similar way, all binary pulsars with two measurable PK parameters yield solutions for their component masses. At present, most of the experimental data on the masses of neutron stars (see Figure 9) come from such timing analyses of binary pulsar systems (36, 37, and references therein).

VI. TESTING FOR GRAVITATIONAL WAVES

If three or more post-Keplerian parameters can be measured for a particular pulsar, the system becomes over-determined, and the extra experimental degrees of freedom transform it into a calibrated laboratory for testing relativistic gravity. Each measurable PK parameter beyond the first two provides an explicit, quantitative test. Because the velocities and gravita-

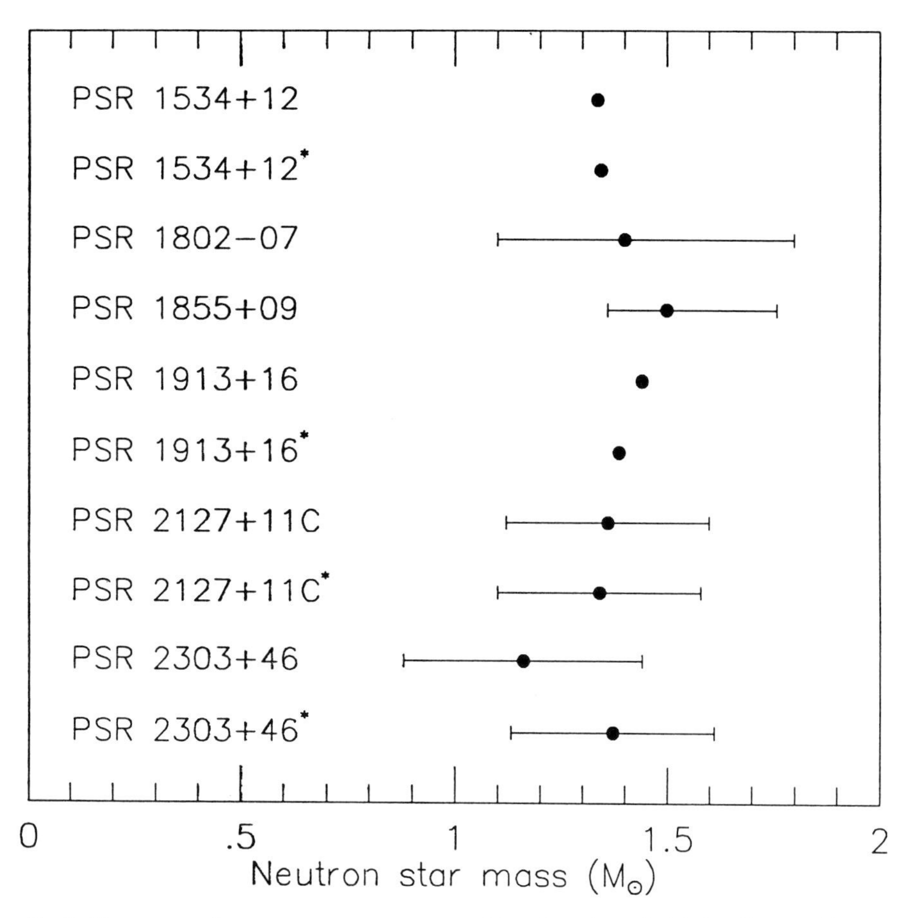

Fig. 9: The masses of ten neutron stars, measured by observing relativistic effects in binary pulsar orbits.

tional energies in a high-mass binary pulsar system can be significantly relativistic, strong-field and radiative effects come into play. Two binary pulsars, PSRs 1913+16 and 1534+12, have now been timed well enough and long enough to yield three or more PK parameters. Each one provides significant tests of gravitation beyond the weak-field, slow-motion limit (32, 38).

PSR 1913+16 has an orbital period $P_b \approx 7.8$ h, eccentricity $e \approx 0.62$ and mass function $f_1 \approx 0.13\ M_\odot$. With the available data quality and time span, the Keplerian orbital parameters are actually determined with fractional accuracies of a few parts per million, or better. In addition, the PK parameters $\dot{\omega}$, γ, and $\dot{P}_b$ are determined with fractional accuracies better than $3 \cdot 10^{-6}$, $5 \cdot 10^{-4}$, and $4 \cdot 10^{-3}$, respectively (25, 39). Within any viable relativistic theory of gravity, the values of $\dot{\omega}$ and γ yield the values of m_1 and m_2 and a corresponding prediction for $\dot{P}_b$ arising from the damping effects of gravitational radiation. At present levels of accuracy, a small kinematic correction (approximately 0.5% of the observed $\dot{P}_b$) must be included to account for accelerations of the solar system and the binary pulsar system in

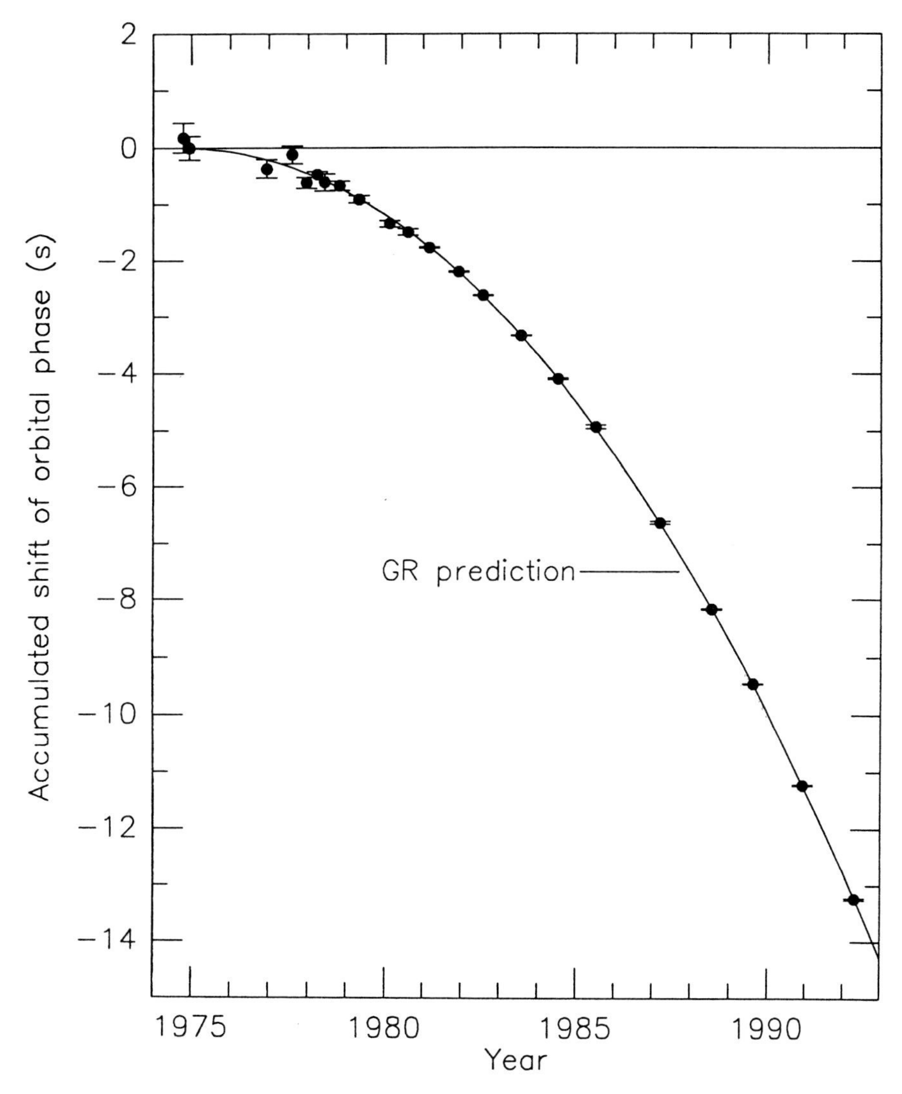

Fig. 10: Accumulated shift of the times of periastron in the PSR 1913+16 system, relative to an assumed orbit with constant period. The parabolic curve represents the general relativistic prediction for energy losses from gravitational radiation.

the Galactic gravitational field (40). After doing so, we find that Einstein's theory passes this extraordinarily stringent test with a fractional accuracy better than 0.4% (see Figures 10 and 11). The clock-comparison experiment for PSR 1913+16 thus provides direct experimental proof that changes in gravity propagate at the speed of light, thereby creating a dissipative mechanism in an orbiting system. It necessarily follows that gravitational radiation exists and has a quadrupolar nature.

PSR 1534+12 was discovered just three years ago, in a survey by Aleksander Wolszczan (41) that again used the huge Arecibo telescope to good advantage. This pulsar promises eventually to surpass the results now avail-

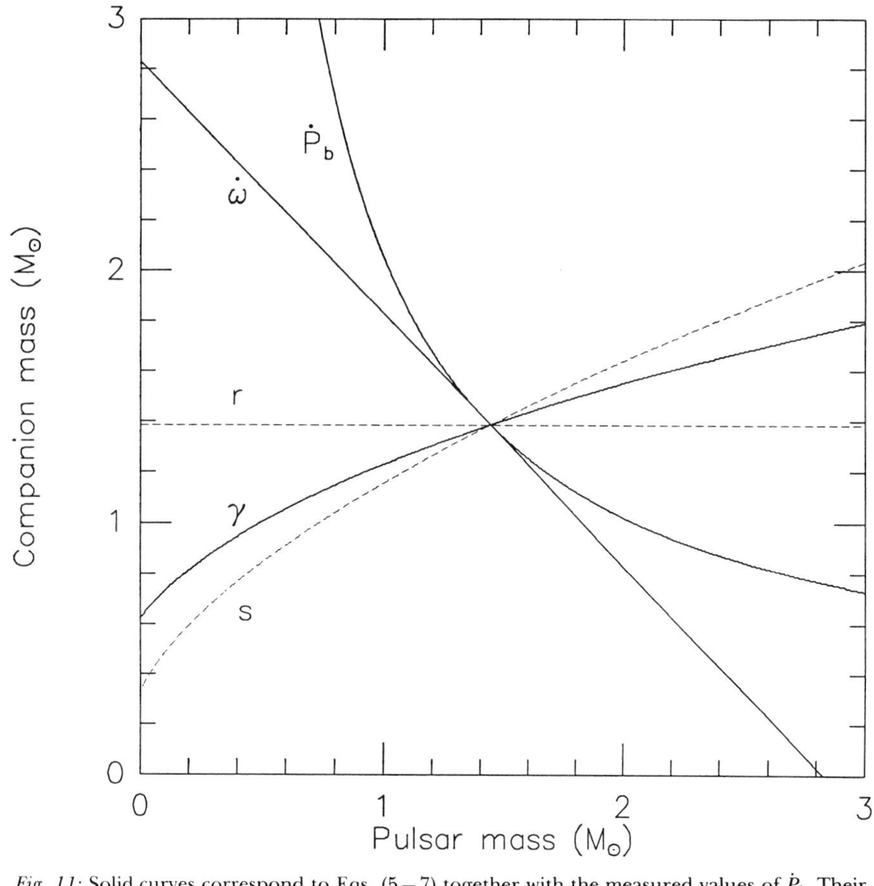

Fig. 11: Solid curves correspond to Eqs. (5 – 7) together with the measured values of $\dot{P}_b$. Their intersection at a single point (within the experimental uncertainty of about 0.35% in $\dot{P}_b$), establishes the existence of gravitational waves. Dashed curves correspond to the *predicted* values of parameters r and s; these quantities should become measurable with a modest improvement in data quality.

able from PSR 1913+16. It has orbital period $P_b \approx 10.1$ h, eccentricity $e \approx 0.27$, and mass function $f_1 \approx 0.31\ M_\odot$. Moreover, with a stronger signal and narrower pulse than PSR 1913+16, its TOAs have considerably smaller measurement uncertainties, around 3 μs for five-minute observations. Results based on 15 months of data (39) have already produced significant measurements of four PK parameters: $\dot{\omega}$, γ, r, and s. In recent work not yet published, Wolszczan and I have measured the orbital decay rate, $\dot{P}_b$, and found it to be in accord with general relativity at about the 20% level. In fact, *all* measured parameters of the PSR 1534+12 system are consistent within general relativity, and it appears that when the full experimental analysis is complete, Einstein's theory will have passed three more very stringent tests under strong-field and radiative conditions.

I do not believe that general relativity necessarily contains the last valid words to be written about the nature of gravity. The theory is not, of course, a quantum theory, and at its most fundamental level the universe appears to obey quantum-mechanical rules. Nevertheless, our experiments with binary pulsars show that, whatever the precise directions of future theoretical work may be, the correct theory of gravity must make predictions that are asymptotically close to those of general relativity over a vast range of classical circumstances.

ACKNOWLEDGMENTS

Russell Hulse and I have many individuals to thank for their important work, both experimental and theoretical, without which our discovery of PSR 1913+16 could not have borne fruit so quickly or so fully. Most notable among these are Roger Blandford, Thibault Damour, Lee Fowler, Peter McCulloch, Joel Weisberg, and the skilled and dedicated technical staff of the Arecibo Observatory.

REFERENCES

1. A. Hewish, S. J. Bell, J. D. H. Pilkington, P. F. Scott, and R. A. Collins. Observation of a rapidly pulsating radio source. *Nature*, 217:709–713, 1968.
2. G. R. Huguenin, J. H. Taylor, L. E. Goad, A. Hartai, G. S. F. Orsten, and A. K. Rodman. New pulsating radio source. *Nature*, 219:576, 1968.
3. T. Gold. Rotating neutron stars as the origin of the pulsating radio sources. *Nature*, 218:731–732, 1968.
4. D. H. Staelin and E. C. Reifenstein, III. Pulsating radio sources near the Crab Nebula. *Science*, 162:1481–1483, 1968.
5. M. I. Large, A. E. Vaughan, and B. Y. Mills. A pulsar supernova association. *Nature*, 220:340–341, 1968.
6. D. W. Richards and J. M. Comella. The period of pulsar NP 0532. *Nature*, 222:551–552, 1969.
7. V. Radhakrishnan and D. J. Cooke. Magnetic poles and the polarization structure of pulsar radiation. *Astrophys. Lett.*, 3:225–229, 1969.
8. P. Goldreich and W. H. Julian. Pulsar electrodynamics. *Astrophys. J.*, 157:869–880, 1969.
9. R. N. Manchester and W. L. Peters. Pulsar parameters from timing observations. *Astrophys. J.*, 173:221–226, 1972.
10 .W. R. Burns and B. G. Clark. Pulsar search teechniques. *Astron. Astrophys.*, 2:280–287, 1969.
11. R. N. Manchester, J. H. Taylor, and G. R. Huguenin. New and improved parameters for twenty-two pulsars. *Nature Phys. Sci.*, 240:74, 1972.
12. J. H. Taylor. A sensitive method for detecting dispersed radio emission. *Astron. Astrophys. Supp. Ser.*, 15:367, 1974.
13. J. H. Taylor, R. N. Manchester, and A. G. Lyne. Catalog of 558 pulsars. *Astrophys. J. Supp. Ser.*, 88:529–568, 1993.
14. F. Camilo. Millisecond pulsar searches. In A. Alpar, editor, *Lives of the Neutron Stars, (NATO ASI Series)*, Dordrecht, 1994. Kluwer.
15. J. H. Taylor. A high sensitivity survey to detect new pulsars. Research proposal submitted to the US National Science Foundation, September, 1972.
16. R. A. Hulse and J. H. Taylor. A high sensitivity pulsar survey. *Astrophys. J. (Letters)*, 191:L59–L61, 1974.

17. R. A. Hulse and J. H. Taylor. Discovery of a pulsar in a binary system. *Astrophys. J.*, 195:L51–L53, 1975.
18. R. A. Hulse and J. H. Taylor. A deep sample of new pulsars and their spatial extent in the galaxy. *Astrophys. J. (Letters)*, 201:L55–L59, 1975.
19. R. A. Hulse. The discovery of the binary pulsar. *Les Prix Nobel 1993*, 58–79. The Nobel Foundation, 1994.
20. J. H. Taylor. Millisecond pulsars: Nature's most stable clocks. *Proc. I. E. E. E.*, 79:1054–1062, 1991.
21. J. H. Taylor, R. A. Hulse, L. A. Fowler, G. E. Gullahorn, and J. M. Rankin. Further observations of the binary pulsar PSR 1913+16. *Astrophys. J.*, 206:L53–L58, 1976.
22. P. M. McCulloch, J. H. Taylor, and J. M. Weisberg. Tests of a new dispersion-removing radiometer on binary pulsar PSR 1913+16. *Astrophys. J. (Letters)*, 227:L133–L137, 1979.
23. J. H. Taylor, L. A. Fowler, and P. M. McCulloch. Measurements of general relativistic effects in the binary pulsar PSR 1913+16. *Nature*, 277:437, 1979.
24. J. H. Taylor and J. M. Weisberg. A new test of general relativity: Gravitational radiation and the binary pulsar PSR 1913+16. *Astrophys. J.*, 253:908–920, 1982.
25. J. H. Taylor and J. M. Weisberg. Further experimental tests of relativistic gravity using the binary pulsar PSR 1913+16. *Astrophys. J.*, 345:434–450, 1989.
26. D. R. Stinebring, V. M. Kaspi, D. J. Nice, M. F. Ryba, J. H. Taylor, S. E. Thorsett, and T. H. Hankins. A flexible data acquisition system for timing pulsars. *Rev. Sci. Instrum.*, 63:3551–3555, 1992.
27. R. Blandford and S. A. Teukolsky. Arrival-time analysis for a pulsar in a binary system. *Astrophys. J.*, 205:580–591, 1976.
28. R. Epstein. The binary pulsar: Post Newtonian timing effects. *Astrophys. J.*, 216:92–100, 1977.
29. M. P. Haugan. Post-Newtonian arrival-time analysis for a pulsar in a binary system. *Astrophys. J.*, 296:1–12, 1985.
30. T. Damour and N. Deruelle. General relativistic celestial mechanics of binary systems. I. The post-Newtonian motion. *Ann. Inst. H. Poincaré (Physique Théorique)*, 43:107–132, 1985.
31. T. Damour and N. Deruelle. General relativistic celestial mechanics of binary systems. II. The post-Newtonian timing formula. *Ann. Inst. H. Poincaré (Physique Théorique)*, 44:263–292, 1986.
32. T. Damour and J. H. Taylor. Strong-field tests of relativistic gravity and binary pulsars. *Phys. Rev. D*, 45:1840–1868, 1992.
33. L. A. Rawley, J. H. Taylor, and M. M. Davis. Fundamental astrometry and millisecond pulsars. *Astrophys. J.*, 326:947–953, 1988.
34. M. F. Ryba and J. H. Taylor. High precision timing of millisecond pulsars. I. Astrometry and masses of the PSR 1855+09 system. *Astrophys. J.*, 371:739–748, 1991.
35. V. M. Kaspi, J. H. Taylor, and M. Ryba. High-precision timing of millisecond pulsars. III. Long-term monitoring of PSRs B1855+09 and B1937+21. *Astrophys. J.*, in press, 1994.
36. J. H. Taylor and R. J. Dewey. Improved parameters for four binary pulsars. *Astrophys. J.*, 332:770–776, 1988.
37. S. E. Thorsett, Z. Arzoumanian, M. M. McKinnon, and J. H. Taylor. The masses of two binary neutron star systems. *Astrophys. J. (Letters)*, 405:L29–L32, 1993.
38. J. H. Taylor, A. Wolszczan, T. Damour, and J. M. Weisberg. Experimental constraints on strong-field relativistic gravity. *Nature*, 355:132–136, 1992.
39. J. H. Taylor. Testing relativistic gravity with binary and millisecond pulsars. In R. J. Gleiser, C. N. Kozameh, and O. M. Moreschi, *editors, General Relativity and Gravitation 1992*, pages 287–294, Bristol, 1993. Institute of Physics Publishing.

40. T. Damour and J. H. Taylor. On the orbital period change of the binary pulsar PSR 1913+16. *Astrophys. J.*, 366:501–511, 1991.
41. A. Wolszczan. A nearby 37.9 ms radio pulsar in a relativistic binary system. *Nature*, 350:688–690, 1991.

Physics 1994

BERTRAM N. BROCKHOUSE

for his development of neutron spectroscopy

and

CLIFFORD G. SHULL

for the development of the neutron diffraction technique

THE NOBEL PRIZE IN PHYSICS

Speech by Professor Carl Nordling of the Royal Swedish Academy of Sciences.
Translation from the Swedish text.

Your Majesties, Your Royal Highnesses, Ladies and Gentlemen,

This year's Nobel Prize in Physics has been awarded to Bertram Brockhouse and Clifford Shull for their pioneering contributions to the development of neutron scattering techniques for the study of liquid and solid matter. In simple terms, one could say that Shull answered the question of where atoms "are," while Brockhouse answered the question of what they "do."

At the end of World War II, research conditions underwent a radical change, especially in the United States. For some years, every single neutron emitted by a radioactive source, produced in an accelerator or released in a nuclear reactor had been employed for one single purpose: to produce the first atomic bomb. Suddenly a major new resource was being placed in the service of peaceful research. Neutrons could henceforth perform other tasks besides splitting atomic nuclei. Words like TOP SECRET were no longer automatically stamped across the cover of each research report.

Researchers had been familiar with neutrons as building blocks in the atomic nucleus for more than a decade. They had also done some thinking and conducted experiments concerning the properties of neutrons as free particles outside the nucleus. For example, they knew that neutrons possessed a dual nature that was characteristic of their tiny world: the ability to behave both as particles and as waves. In their latter guise, neutrons had been reflected against the atomic planes of a crystal in the same way X-rays previously had.

This provided a hint that some day, neutrons might become a tool for studying the microstructure of matter at the atomic level. The door was already ajar, but had not yet been opened wide.

Brockhouse and Shull followed their own individual strategies, both with the aim of gaining new knowledge about liquid and solid materials, otherwise called "condensed matter."

Shull took advantage of the fact that the wavelength of neutrons from a reactor may be roughly equal to the distance between the atoms in a solid body or a liquid. When the neutrons bounce against atomic nuclei, they do not lose energy, but their scattering is concentrated in directions that are determined by the structure in which the atoms are arranged. Shull revealed that neutrons could answer questions that the X-ray diffraction method had failed to answer, such as where the atoms of the light element hydrogen are located in an ice crystal.

Another breakthrough concerned magnetic structures. Neutrons themselves are small magnets and can interact very efficiently with the atoms in a magnetic material. Shull demonstrated how neutrons can reveal the magnetic properties of metals and alloys. The X-ray method had been powerless to accomplish this task as well.

While Shull was studying elastic neutron scattering, that is, scattering that occurs without energy changes, Brockhouse was concentrating on inelastic scattering. In the latter, neutrons lose part of their energy to the material or pick up energy from it.

Brockhouse designed ingenious instruments with which he managed to record the energy spectrum of the scattered neutrons. This enabled him to gather new information about such phenomena as atomic vibrations in crystals, diffusion movements in liquids and fluctuations in magnetic material. As a consequence, the study of these types of phenomena underwent a renaissance.

Over the years since Brockhouse and Shull made the contributions for which they are now being awarded the Nobel Prize, their methods have found widespread applications. Thousands of researchers are using neutron scattering to study the structure and dynamics of the new ceramic superconductors, molecule movements on surfaces for catalytic exhaust emission control, the interaction between proteins and the genetic material of viruses, the connection between the structure and elastic properties of polymers, the rapidly fading memory of the atomic structure of a metallic melt and much more. The pioneers of this broad field of research are the recipients of this year's Nobel Prize in Physics.

Professor Brockhouse, Professor Shull,
You have been awarded the 1994 Nobel Prize in Physics for your pioneering contributions to the development of neutron scattering in condensed matter research. It is my privilege to convey to you the heartiest congratulations of the Royal Swedish Academy of Sciences, and I now ask you to receive the Prize from the hands of His Majesty the King.

Bertram N. Brockhouse

BERTRAM N. BROCKHOUSE

It appears that I was born in hospital in Lethbridge, Alberta, Canada on July 15, 1918. My first memories are of a farm near Milk River where I lived with my mother and father and my sister, Alice Evelyn, and a variety of farm and domestic animals. My father, Israel Bertram Brockhouse, had homesteaded with other members of his family in 1910. He had spent his years to that time in the United States after being brought to this continent at two years of age from the family's native Yorkshire. My mother, Mable Emily (Neville) Brockhouse had grown up in Illinois, the product of uncounted generations of North American English people. As the years went on there were two other children born: Robert Paul, who died in infancy, and Gordon Edgar who became much later a railroad civil engineer. In the winter of 1926 – 27 our family moved to Vancouver B.C. and it was in that city my sister and brother and I grew up.

My sister entered the school system in a normal way. But I had been a somewhat nominal attendee of the one-room elementary school a couple of miles from our farm and my preparation for the system was somewhat mixed. I must have learned to read and to do simple arithmetic at a very early age because I cannot remember ever learning these subjects. But in other ways I was much behind my potential classmates. But the fine Vancouver schools I attended (Central and then Lord Roberts elementary schools and King George High School – and the Sunday School of St. John's United Church) soon took care of this. So I had what I believe to be a good basic education, except for social and organizational defects probably arising from the facts that I found school work easy and that I was younger than most of my classmates.

There were other people of course who had influence on me. These included my two aunts: Edith (Neville) Murphy in Chicago and Maude (Brockhouse) Smith in western Canada. My older cousin Wilbert B. Smith may have inspired an early interest in radio technology.

Our family finances were somewhat precarious so I carried newspapers for most of my teens. But the Great Depression made things worse and in 1935 our family moved by train to Chicago in the hope of bettering the situation. I had completed High School by this time and took some evening courses at Central YMCA College (now Roosevelt University). I was interested in the technical aspects of radios and learned to repair and design and build them. This and my facility with mathematics was, I suppose, what pointed me eventually in the direction of physics. For part of our time in Chicago I worked as a lab assistant in a small electronic firm, Aubert Controls Corporation. But the company failed in the recession of 1937. In 1938 our family decided to return to Vancouver and we drove across the continent, all of us I think enjoying the experience.

In Chicago I had begun to repair radios as a small business and I continued this in Vancouver. My parents ran a small grocery store but neither enterprise was really successful. I had always been interested in politics but now I began to take part as an active member of the leftist party of the era, the CCF. My adherance to the CCF continued for many years, in fact until I became an employee of the Dominion Government in the shape of the Chalk River Laboratory. (I understood then (and still do) that there is something dishonourable in a democratic society for a Government employee being other than politically neutral). I was profoundly anti-totalitarian and hence anti-communist so that when World War II erupted I was motivated from many sides to join the military. On September 26, 1939 I enlisted in the Royal Canadian Navy with the design of becoming a Radio Telegrapher. In the event I spent some months at sea as a seaman and ASDIC operator but spent most of my six years in the Navy servicing ASDIC equipment at a shore base. In 1944 I was enrolled in a six-month course in Electrical Engineering at Nova Scotia Technical College and then as a newly-minted Electrical Sub-Lieutenant assigned to the test facilities at the National Research Council in Ottawa. It was there that I met Doris Miller, the girl who later became my wife.

The war having ended, in late August 1945 I was drafted home to Vancouver and was discharged from the Navy on September 11 1945, under the principle "first in – first out". The Department of Veterans' Affairs was ready to supply finances for either a small land-holding or for training or education. Thus the way was clear for me to start immediately at the University of British Columbia. My preparation was such that the obvious choices for my course of study was either Electrical Engineering or Physics and I chose to enroll in Physics and Mathematics. I did very well in my first year, actually winning a scholarship. The university life was probably not typical because many of us were older than would normally have been the case. It was not all study, I operated also a (very) small business which eased our financial problems and I owned a motorcycle for transportation and enjoyment.

In the summer of 1946 after taking a summer class for extra credit, I took a vacation on my motorcycle, going all the way to Ottawa via Chicago. This was probably a decisive step in my life because I took up with Dorie again. With time short I returned with my motorcycle by train to Vancouver. Just before Christmas of 1946 my father died. He had long been troubled with a heart condition so his death was not a surprise. In the spring of the year Alice married so our family was now considerably changed. I had received some University credit for my irregular courses in mathematics and electricity and together with overload credits I was able to complete my B.A. program in

April. I had been offered a summer job in the Nationel Research Council laboratory (the electrical standards section) so off I went to Ottawa again. There Dorie and I became engaged to be married.

It had been arranged that I should return to Vancouver to take a Master's degree course but instead I went to the Low Temperature Laboratory of the University of Toronto. This was one of the two Universities in Canada to offer Ph. D. programs at that time (the other was McGill in Montreal). Being already 29 years of age I was very anxious to embark on my physics career. Furthermore, partly no doubt for financial reasons, DVA was very keen that I do my studies in Canada. So I started work under the guidance of Professors Hugh Grayson-Smith and James Reekie on the effects of stress and temperature on ferro-magnetism and finished a Master's program in the then normal period of eight months. In May, Dorie and I were married in the village of Kirkfield, the old home of her family. For the remainder of the summer we lived in Ottawa, Dorie continuing as a film technician at the National Film Board while I worked as a summer student in the acoustics section of the National Research Council. The more passive part of my education was now complete. The instruction via course-work which I received at UBC and Toronto was probably as good as I could reasonably have expected. Certainly I remember almost all the teachers and courses with fondness. Partly because my mind was "already formed" I suppose, I did not become comfortable with Quantum Mechanics and indeed never did so. The classical nature of the small researches I performed contributed to what was probably an "old-fashioned outlook" even at the time. And now I was forced to assume full responsibility for my future – and the future of my new family.

The Low-Temperature Laboratory at Toronto was long-established and reasonably well-equipped. But at this point my supervisors both left to assume more senior positions at other institutions. Furthermore the third faculty member in the Lab also left. So I was left essentially unsupervised and should also have moved – except that we were now expecting the birth of our first child. But happily, as we thought, Sir Edward Bullard, an expert in earth-magnetism, was coming to head the Department – and to assume direction of my thesis work. If he had stayed for longer than he did then possibly I would have changed my field and worked on the earth-magnetism problems then very current and in which I had some interest. But he left to assume a high position in the U.K. so ultimately I had to do the best I could while receiving every possible help from the Department.

My thesis subject was a contribution to Solid State physics which involved experiments at both low and high temperatures. There were a few books on the general subject, two excellent ones being by Frederick Seitz and by N.F. Mott and H. Jones. These I to a considerable extent devoured. I had had lectures on the subject from Grayson-Smith and had a small correspondence with him. I had courses in Thermodynamics, Statistical Mechanics and Theory of Errors. I took a course on Nuclear Theory from my friend Melvin Preston, who was then at Toronto. So I was not too badly prepared in a general way for work on the periphery of Nuclear Energy, when the chance to work at Chalk River was offered to me.

In August 1950 I went up to Deep River, in the van carrying our belongings, while Dorie (and baby Ann) stayed with her parents in a cottage on Balsam Lake near Kirkfield. There I met Don Hurst in whose (neutron physics) group I was to work and saw the house on Hillcrest Ave which was assigned to us. In a short while Dorie (and Gordon-soon-to-be) and Ann joined me. There was still some work to do on my thesis so I would be very busy for the next months. But in October Gordie was born and I passed my Ph. D. exam and we were set for the next period of our lives.

We had originally thought of staying for only a few years and then going on, probably to a University. In the event we stayed for twelve years and four more children. As I progressed we moved (twice) to a better house as was the custom. Despite my long and irregular hours each of us had a social life and one together and we have kept in touch with some of our acquaintance then to this day. Since the work I did then represents a major part of the content of my lecture I will here be brief; I have reviewed it elsewhere – the major advance at this time in early 1951 was the realization that phonons could be studied by studying inelastic scattering and that evocative experiments to do so might be feasible at Chalk River.

The first actual experiments studied the scattering of neutrons by highly absorbing elements, in the process verifying the famous Breit-Wigner formula. This work (on scatterers Cd, Sm and Gd) was done in collaboration with Myer Bloom and D. G. Hurst and was published in Physical Review (1951) and in the Canadian Journal of Research (1953). The apparatus was later much modified and used to study the inelastic scattering from several materials (Aluminium, Graphite and Diamond) by absorption methods. This was the first quantitative experiment in slow neutron spectroscopy and was published in Physical Review. Other experiments by absorption methods were done about the same time at Harwell by R. D. Lowde and P. A. Egelstaff; that by Ray Lowde was particularly significant as it went far to establish the concept "spin wave" on a microscopic basis.

Preparations were underway to attempt proper (differential) studies of inelastic scattering and some almost futile attempts had been made, when our work was terminated by an accident to the NRX high flux reactor which was the source of the neutrons we used. This occurred in November, 1952 and I did not resume actual experiments at NRX until the summer of 1954. Fortunately, I was invited to go to Brookhaven National Laboratory and was able to spend most of one year there with my family, returning to Deep River in February, 1954. The time was very profitable for me, I worked on several experiments, with collaborators and without. But I did not do any spectroscopic work though I met Donald Hughes and Harry Palevsky, now also thinking about inelastic scattering and in particular thinking about the "Cold Neutron" or (Beryllium) Filter-Chopper method. And I met Leon Van Hove and learned about the new generalized (time-dependent) correlations which Noel K. Pope and I were later to put to good use.

After NRX was available to us again in August 1954, things progressed rapidly. Because of the efforts of David G. Henshaw and Jack Freeborn, we had metal monochromators of greatly improved efficiency compared with the NaCl crystals which we were using in 1952. Alec T. Stewart was rapidly getting the Be/Pb Filter-Chopper apparatus together and the primitive Triple-Axis spectrometer was functioning. So I was able to present a paper with substantial (if primitive) results at the New York meeting of the American Physical Society at the end of January, 1955. Publications followed soon after, in Physical Review and in the Canadian Journal of Research.

In 1956 we were able to complete the first true Triple-Axis crystal spectrometer, though only for operation at constant incoming energy. The flexibility of operation and the accuracy of the results were both greatly improved. The "Constant Q Method" was invented in 1958 and at about the same time a new apparatus allowing operation with variable incoming energy was installed at the new high-flux reactor NRU. (Ed Glaser and William McAlphin played crucial roles in these developments.) With the considerable improvements in both the neutron flux and the operating conditions afforded by NRU the subject entered a new phase in 1959. The Triple-Axis spectrometer thereby reached nearly full development. Visitors from other countries were now arriving to spend time working in the group. (The first such visitor was P. K. Iyengar from India who with several others became a life-long friend.) From about 1958 on the interest shifted, from the neutron physics and the methods and the validity of the theory, to the specific results and interpretation for the specific speciment material.

In 1956 also Alec Stewart completed the Filter-Chopper apparatus. This was an equipment similar in general to that of Hughes and Palevsky; it was used in experiments on Aluminium and Vanadium, both chosen for the same good technical reasons that others chose to work on them. When Stewart left to become a professor at Dalhousie University I converted the instrument to the first "Rotating Crystal Spectrometer" – a bad choice of name as it should have been termed "Spinning Crystal". This instrument was used principally to study liquids and polycrystals, as was its improved successor at the NRU reactor.

Three other major technological initiatives were taken. Filters of (large, perfect) single-crystals (quartz), preferably cooled to low temperatures, enabled major improvement in the ratio of slow neutrons to fast in the primary beam and thus in the signal to background ratio. The "Beryllium-Detector" method was developed by enabling the Triple-Axis spectrometer to accept Beryllium polycrystalline filters in the scattered beam and thus, with incoming neutrons of variable energy, to get energy distributions in a different and sometimes advantageous manner – an inverse of the Filter-Chopper method. Finally profitable uses of the new material, pyrolitic graphite, were found – as filters and as crude monochromators.

As time went on I began to receive invitations to attend Conferences and colloquia. In 1957 I made my first trip to England and Europe. Aside from several seminars, I gave a paper in September at a Conference on the Physics and Chemistry of Liquids; held in Varenna on Lago di Como in Italy. My last stop was at a gathering in Stockholm of neutron scattering people. After giving my paper on the first day I became ill with "flu" and spent the next few days of my trip to Europe in hospital. Nevertheless the trip was very inspirational and rewarding. In October 1960, this time accompanied by Dorie, I made another trip to Europe, and gave papers at two IAEA Conferences in Vienna. One of these was the first of the IAEA Conferences on Inelastic Scattering that played such a large role in the development of the subject.

In 1958 our group was joined by A.D.B. (David) Woods, who from then on was my closest collaborator. Numerous people spent periods of time in the group. Of these I must mention William Cochran who collobarated in the project to study the lattice vibrations in alkali halide crystals and in the course of this work developed his well-known "shell model" for the atoms in these and other crystals. Following this, his student from Cambridge University, Roger A. Cowley, commenced his own long association with the group. In 1961 Gerald Dolling arrived after studies at Cambridge and Harwell (with G. L. Squires); he is the only person among those mentioned who is still active in the group.

Other colleagues at Chalk River and visistors there were important to my program. These included: I. L. (Dick) Fowler, Harris McCrady, Walther Woytowich, C. W. Crawford, C. E. L. Gingell, William Howell, G. R. DeMille, Guiseppe Caglioti, T. Arase, R. G. Johnson, K. R. Rao, M. Sakamoto, Hiroshi Watanabe, Leo N. Becka, Roger N. Sinclair, B. A. Dasannacharya, R. H. (Bob) March, A. E. (Ted) Dixon, R. Sherman Weaver, J. Bergsma...

In 1962 I took up a position as Professor of Physics at McMaster University, in Hamilton, Ontario. The research program that I had embarked on eleven years before had been successful beyond expectations and the field was becoming well established. For over fifteen years it had been my intention to take up a University career and in my mid-forties it seemed that "now" was the time if I were to do so. McMaster had a "swimming-pool" reactor which promised to make the transition easier on the research side. For social reasons I preferred not to join a mega-university or live in a mega-city, partly because I thought that it would be better for our family of six children. Dorie was supportive of these ideas. So off we went in the summer of 1962, first to a house in Dundas and soon after to the house in Ancaster in which we still live.

Chalk River had been very good to us. And now the Laboratory facilitated our transfer and encouraged my plans to continue a research program based at McMaster and to use the reactor there for training students and for preliminary work on experiments to be carried out at Chalk River. This arrangement was I think very successful all through the 1960s and early '70s and indeed has been carried on by others since that time. At McMaster a talented group of students put together a neutron diffractometer and a triple-axis instrument and these were available from 1965 on – and indeed are still in use. For the first years we used existing equipment at Chalk River but about

1971 we installed our own spectrometer at NRU and the smaller group now working with me used it (as did others) until I completely left neutron scattering about 1979.

Deep River was also good to us. Five of our six children were born in Deep River Hospital. (Gordon Peter, Ian Bertram, James Christopher, Alice Elizabeth and Charles Leslie.) Our contacts with friends made then have remained deep. But there was one matter for distress – in babyhood Jamie developed hyperactive and autistic behaviour and in 1961 he was placed in Smith Falls Hospital School where he remained until, somewhat improved, we brought him home to Ancaster in 1967. He was sent to special schools in Hamilton; since then he has worked in a sheltered workshop. Of late years he has lived with other afflicted persons in a supervised apartment. Our other five children have all gone on to successful careers; Charles, a molecular biologist, is the only scientist among them. We now have eight grandchildren in four families.

At McMaster I lived the normal life of a Professor of Physics. Each year I usually taught two courses (mostly Solid State Physics, Thermodynamics and Statistical Mechanics) and carried out the other duties required of me. Eleven people won wheir Ph. D. degrees under my supervision: S. H. Chen, J. M. Rowe, E. C. Svensson, S. C. Ng, A. P. Miller, E. D. Hallman, J. R. D. Copley, A. P. Roy, W. A. Kamitakahara, H. C. Teh, A. Larose. About half of them found their careers in neutron scattering. The research of the group consisted of studies of the phonons in crystals and their temperature behaviour, especially in single crystals of metallic alloys. There were also several Master's projects, one of which should be mentioned: the highly quantitative study by R. R. Dymond of the reflective behaviour of maltreated copper monochromators. The contributions of several other men should be mentioned, including G. A. DeWit, William Scott, James Couper, E. Roger Cowley, A. K. Pant, Jake Vanderwal and David Macdonald.

But my greatest debt is to my wife of 46 years and my family, whose support and encouragement were indispensable and total. And following this, my colleagues and I owe gratitude to the technologists who engineered and maintained the reactors which provided the neutrons employed in the work – and to Don Hurst who introduced me to the subject – and to the National Research Council of Canada, who supported the program at McMaster over many years – and, finally, to the people of Canada, who supported all these and us.

From 1960 on I suffered, at intervals of a few years, serious health problems of several varieties. These were kept under control by our medical allies and by the support of Dorie and our families. My work was not affected much in formal ways though undoubtedly some apsects did suffer. Throughout my career my father-in-law (Sidney L. Miller) maintained a cottage on Balsam Lake, north of Toronto; this was a considerable blessing for all of us. In addition we did a little camping from time to time, until I developed a bad back. And music – consert, opera, records – have always been part of our life.

During the 1970s I gradually realigned my intellectual interests. One avenue I explored was what might loosely be termed "philosophy of physics". Another (intersecting) route was concerned with energy supply and the economics and ethics thereof. And there were others. In my explorations I entertained the hope that I would find some interesting niche in which to work. But I also realized the extreme importance of reaching general points of view, if this were at all possible. In this quest I struggled with new descriptions of the furniture of the world. Not much of what I sought was found and not much of that was made public – though I did give some seminars and some talks to service clubs and the like. Perhaps the new impetus to action, given by the amazing event of the Nobel Prize and its accompaniments, will move me on to produce something more well-defined.

VITA

BERTRAM NEVILLE BROCKHOUSE: Born July 15,1918 at Lethbridge, Alberta, Canada • Son of Israel Bertram and Mable Emily (Neville) Brockhouse • Educated: King George High School, Vancouver, B.C. • Served in the Royal Canadian Navy 1939–1945 • B.A. 1947 (Mathematics and physics) University of British Columbia • M.A. 1948 and Ph.D. 1950 (Physics: Solid State) University of Toronto • Married to Doris Isobel Mary Miller 1948 • Research Officer, Chalk River Laboratory, 1950–1959 • Seconded for 10 months to Brookhaven National Laboratory 1953–54 • Head, Neutron Physics Branch CRNL, 1960–1962 • Member of Council, Canadian Association of Physicists, 1960–61 • Fellow, Royal Society of Canada, 1962 • Oliver S. Buckley Prize of the American Physical Society, 1962 • Professor of Physics, McMaster University, 1962–1984 • Duddell Medal and Prize of the Institute of Physics, 1963 • Fellow, Royal Society of London, 1965 • Chairman, Department of Physics, McMaster University, 1967–70 • Medal for Achievement in Physics, Canadian Association of Physics, 1967 • Centennial Medal of Canada, 1967 • Doctor of Science, Honoris Causa, University of Waterloo, 1969 • Member of Council, American Physical Society, 1969–1973 • Sabbatical year (1970–71 - Guggenheim Fellowship - BNL, ORNL in USA, AERE Harwell, England) • Tory Medal of the Royal Society of Canada, 1973 • Memberships in societies for History and Philosophy of Science • Officer, The Order of Canada, 1982 • Retired in 1984. Professor Emeritus of Physics, McMaster University • Doctor of Science, Honoris Causa, McMaster University, 1984 • Foreign Member, Royal Swedish Academy of Sciences, 1984 • Honorary Foreign Member, American Academy of Arts and Sciences, 1990 • Nobel Prize in Physics, 1994 (with C.G. Shull) • some 90 publications, mostly in neutron physics • six children, eight grandchildren.

SLOW NEUTRON SPECTROSCOPY AND THE GRAND ATLAS OF THE PHYSICAL WORLD

Nobel Lecture, December 8, 1994

by

Bertram N. Brockhouse

Department of Physics, McMaster University, Hamilton, ONT, L8S 4M1, Canada

On October 12, 1994, telephone communications from Stockholm ensured that I would have the privilege of giving a Nobel Lecture, for which I must thank all those involved in arranging for this great and surprising event of my life. But I had to go back some thirty to forty-five years, first in memory and then in the entropy of my files and library. The lecture was given on December 8, 1994 (and subsequently on several occasions). This written version covers the same ground, though sometimes in different order, with a little additional material from my preparations, which time prevented including in the spoken lecture.

In August 1950, when I joined the Physics Division of the Chalk River Nuclear Laboratory, it was 18 years since the identification of the neutron as being emitted in some radioactive processes and 14 years since verification of the supposition that neutrons would exhibit wave-particle duality. A considerable body of theory was available in the open literature – which was still small enough so that one person could have read all that was then available. There had been significant measurements of total cross sections using neutron beams produced by cyclotrons. Self-sustaining reactors employing the neutron-induced fission of uranium had been demonstrated and full-scale models which emitted potent beams of both fast and slow neutrons had been constructed. Slow neutron beams from reactors were already in use at several laboratories (including Chalk River) for studies of crystals and other forms of matter. The works of E.O. Wollan and C.G. Shull at the Oak Ridge National Laboratory were particularly significant because they included the first studies of a number of phenomena and because they already presented values of scattering power for neutrons of a substantial number of elements. This history is discussed in the lecture of my senior colleague and co-winner of the 1994 Nobel Prize in Physics, Clifford Shull.

To 1951, some studies had been made of the elastic scattering of monochromatic slow neutrons by specimens in the form of powdered crystals (the neutron analog of Debye-Scherrer patterns for X-rays), which led to improved crystallographic understanding of the substances involved. And there were a few studies of the angular distribution of slow neutrons scattered by certain specimens in the liquid or gaseous form. Most importantly, magnetic scattering of neutrons, by crystalline substances which contained atoms with magnetic moments, had been demonstrated and some simple magnetic structures determined. But the thermal diffuse scattering, expected from theory, had been studied only as the total cross section for incident neutrons of long wavelength (beyond the possibility of Bragg scattering). No spectrometric measurements had been made, though there existed relevant theory and the phenomenon of neutron slowing-down (moderation) gives assurance that energy changes on single-scattering must occur.

Fig. 1: The main floor of the NRX Reactor at the Chalk River Laboratory about 1950. The powder diffractometer constructed by Donald Hurst and associates is visible near the centre of the photograph. Most of the other equipment is concerned with nuclear physics or with the physics of the neutron itself. For reasons of space, each apparatus is located at the end of a long tube. (AECL photo)

Fig. 1 shows the main floor of the hall at Chalk River with the NRX reactor ("pile"), a very powerful one for the times. A variety of experiments were underway, mostly in nuclear physics but the double-axis powder diffractometer, constructed by Dr. Donald Hurst and associates, is visible near the centre of the photograph. The large size of the equipment is enjoined by the necessity of shielding against the fast neutrons which accompany the slow neutron beams – one's own beam and one's neighbours – for low experimental background as well as for biological protection.

It is necessary to remember that many things which a researcher of the present would take for granted were then not available – not available on the market or not available at all. Counting electronics were still under development in-house, recording apparatus was mostly analog and of limited availability as to kind. The first digital computer at Chalk River was delivered in 1954 and occupied a large room with heat-generating electronics (no siliconics!). The single crystals available were mostly of natural origin, though large ingots of NaCl and other alkali halides were produced for use in infrared spectrometry and were available commercially. But opposed to all this sort of thing, was the remarkable openness of the possibilities available. If an experiment could be envisaged, it was likely new or at least repeating it would not be merely trivial.

A. The NEUTRON and its APPLICATIONS

In 1982 a Conference was held at Cambridge in honour of the 50th Anniversary of the discovery by James Chadwick of the free neutron. The Proceedings [A] deal with the numerous consequences of this event and their histories. Beyond those military consequences so important in the media and beyond the applications to the public energy supply (Nuclear Energy), to Nuclear Medicine and to Nuclear Physics and Chemistry, there have come into existence a number of new scientific fields.

These fields involve neutrons produced by nuclear reactions (as "fast" neutrons) with energies usually millions of electron volts (MeV) and then allowed to "slow down" by repeated collisions with atomic nuclei of small atomic weight ("moderation") to a near-equilibrium distribution at a temperature at little above that of the moderator, often water at a temperature somewhat above room temperature. A small tube through the reactor shielding admits a beam of the fast, intermediate and slow neutrons into the experimental hall.

After dealing with the undesired fast neutrons, well-defined beams of the "slow neutrons" are then employed in a variety of ways to study the properties of material specimens placed in the neutron beam (or in subsidiary neutron beams derived from the first). The fields of interest here are known generically by the terms (slow) neutron scattering or neutron diffraction. The scattering may be elastic or inelastic; in the latter case the experiments may involve measurements of the energy changes experienced by the neutrons when scattered by the specimen, in which case the term (slow) neutron spectrometry is often used. When the incident neutrons have energies well below what they would have if in equilibrium at room temperature (0.025 eV) they are often termed "cold neutrons".

Slow neutrons have energies from millivolts (meV) to tenths of an eV and by the same token have wavelengths from 10 Angstroms or more to tenths of an Angstrom. This fact, that neutrons have, simultaneously, energies of the order of the characteristic quantum energies in condensed matter and wavelengths of the same order as inter-atomic dimensions, is of the first importance to the new subjects. There are several other favourable circumstances:

the velocity of the neutrons (hundreds of metres/sec) is suitable for mechanical measurement. For many materials neutrons penetrate deeply before being captured or scattered; this allows many things to be done which are not possible with other radiations.

The Bibliography gives ample access to the various branches of the subject: neutron optics, neutron diffraction and neutron spectrometry, all possibly involving nuclear and/or magnetic scatterers. Reference [B] "Fifty Years of Neutron Diffraction", has many articles from several points of view, by 35 authors (including Dr. Shull and myself) as well as reprints of the 1936 papers of the very first workers in the field. A much larger bibliography compiled up to 1974, by Larose and Vanderwal [Q], shows the remarkable growth in the number of publications in the subject per year, from the 50 or so in the 1950s to about 800 by 1971 and no doubt many more in recent years.

Thus the fields have grown to the extent that one person, or even a committee, can no longer be fully consonant of what has been done, of what is in the literature. For the larger literature of Science in general or Physics in general this is true *a fortiori,* no one – no one at all – can be sure of the non-existence of some particular results, relevant to his projects. In general this is insurmountable but Physics moves step-by-step with Technology into the future; it is the availability or non-availability of technologies in the open market which defines the "present" or "state of the art" of research. Neutron Spectrometry could not flourish until large neutron sources were for some reason available. The researches of Dr. Shull and his colleagues, like those of myself and mine, were by-products of programs set in motion for other reasons.

The GRAND ATLAS

This brings us to the Grand Atlas, viz, the scientistic literature of the state-of-the-art present. There can be histories of the Grand Atlas, but only historical artifacts and theories of history are contained within it. The Grand Atlas includes the "petit atlas" – earthly geography and geology in their most reliable current scientistic guises – and the scientistic, currently accepted, "recipes" of physics and chemistry. Thus the Grand Atlas comprises "maps" of the world we live in, metaphorical maps which just might prove to be metaphysical, maps that link percepts (which can be thought of as metaphorical geography) with other percepts, by means of theory (which involves concepts and which can be thought of as metaphorical geology). The percepts are sense-data: meter-readings, features in a photograph and the like. The concepts are names involved in the apposite theories. But it does not do to forget that percepts are themselves theory-based, that beyond the photograph is the theory of photography – and of the photographer.

The observations (like the theories) are made by reputed scientists; the observers (like the theoreticians) have done their work openly and reported it in ways that permit repetition. The maps of the Grand Atlas can be checked by other scientists, they are scientistic – purportedly backed by men who have a right to an opinion on the matter. Whether the maps are to be called "scientific facts" is to be known only after-the-fact, a posteriori, by users of the maps. Similarly, whether a particular map is to be thought of as pertaining to metaphysical reality (or even to actuality now) is a religious and philosophical matter.

For strictly scientific or technological purposes all this is irrelevant. On a pragmatic view, as on a religious view, theory and concepts are held in faith. On the pragmatic view the only thing that matters is that the theory is efficacious, that it "works" and that the necessary preliminaries and side issues do not cost too much in time and effort. Beyond that, theory and concepts go to constitute a language in which the scientistic matters at issue can be formulated and discussed.

Of course we would like much more than that. We would like to know whether it is possible to say of "Alberta 100 million B.C.": "Here be dinosaurs." or, at least, "Here were dinosaurs.", to say of an aluminium crystal: "Here be Al atoms.", to say of the theoretical nuclei of conceptual Al atoms: "Here be protons and neutrons.", to say of the mathematical Al crystal lattice: "Here be phonons." We would like to be certain of what the Grand Atlas ultimately means. And of course, once obtained, we might not like that knowledge at all. As it is, each of us takes his own scientismic view of these matters, a more or less conscious aspect of his religious views on "the nature of things".

At a given epoch of the "state-of-the-art" there are applications visible – technological or scientific applications – and also perhaps moral implications which go to forbid or enjoin them. And there can be metaphors visible, which modelled upon, may ultimately find places as theory held in faith, in the Grand Atlas. So that metaphors too are to be watched for their moral implications; nuclear fission, nuclear fusion are examples. Might it not be better that these notions never have been thought?

Thus concepts which are not immediately testable by experience – which are not themselves scientific – remain always questionable. At any given epoch Physics are true by definition; if not true now, the Physics in question never were true. The same is true generally of "the scientific". In the world to which the Grand Atlas applies, there is an enduring tension: additional evidence increases the reasonableness of accepting the concepts as actual entities or even as moral, not merely mental, realities – but the burden of proof can always be shifted to the opposite side. Observation or experimentation can suggest an addition or modification to a theory or can support an existing implementation of the theory. Experiment can refute or falsify such an implementation, but a new implementation may be found which agrees sufficiently with experiment to be acceptable. At a deep enough level in the intellectual structure, a theory is very difficult to refute.

SLOW NEUTRON SPECTROMETRY

Theories of the physics of condensed matter involve the most basic aspects of modern physics: the principles of conservation (energy, linear momentum etc.), the chemical elements in various ionic forms, electrons, neutrons, quantum mechanics. Implementations for a particular substance in a particular setting, usually involve drastic approximations if the required quantum statistical calculations are to be possible. Happily, because the nuclear and magnetic interactions between the neutron and atom are (in some sense) weak, the very good "first Born approximation" is applicable, and the neutrons are effectively "decoupled" from the dynamics of the scattering system which can be considered in isolation. The neutron, in being scattered, "causes" transitions between the quantum states of the scattering system but does not change the states.

In general the scattering system is modelled, with assistance of experimental information about it. For example, the chemical constitution would be taken as given. An aluminium crystal would be taken to have the face-centred cubic (FCC) structure indicated from X-ray or neutron diffraction; theoretical calculation would be unlikely to produce this *ab initio.* The dynamics of the so-defined system could be studied a priori by simple models, or could be studied using a phenomenological theory with numerous parameters to be fitted. The Born-von Kármán theory [O] of crystal lattice vibrations can host either approach, at least to some degree. From the model, macroscopic properties (optical, thermodynamic and the like) can be calculated and compared with experiment. Until the advent of neutron spectroscopy, only qualitative comparison was normally possible – the "distance" between the concepts at the atomic level and the percepts at the macroscopic level was simply too great.

Normally neutron energies (with their associated momenta) are inferred from their wavelengths as determined by diffraction from a single crystal through the Bragg Law, or from their times of flight over a known distance. In each case the energy is proportional to the inverse square of the measured quantity. Thus, at the theoretical level, spectrometers based on the two methods have the same law of dispersion. (Any measurement of the same spectrum by the two methods is implicitly a test of quantum mechanics.) However, crystal spectrometers have additional dispersion from the Bragg transformation from angle of reflection to wavelength. The two methods are very different in their technical aspects and the associated experimental problems.

The first method considered at Chalk River (in 1951) and (I think) elsewhere, was to employ time-of-flight to measure both incoming and outgoing neutrons. The "Double-Chopper" method was to employ electrically phased choppers to select the energy of the incoming neutrons and the time-of-flight from the second chopper to the detector to infer the energy of the scattered neutrons. (Bernard Jacrot, at Saclay, built and employed for a time [26] a double-chopper consisting of two large wheels on a rotating shaft.) It was soon apparent that development of a competent Double-Chopper instrument would be technologically demanding. So at Chalk River attention shifted to the use of crystal diffraction for both energy measurements – the "Triple-Axis" neutron crystal spectrometer which will be discussed here later. The Double-Chopper later received extensive development elsewhere, particularly at the hands of Peter Egelstaff at Harwell. Mixed systems were also developed, of which the "Rotating Crystal" spectrometer [43] is probably the best known. This system employs a spinning crystal as initial monochromator and measures the spectrum by time of flight.

There is also an historically important method – the "Filter-Chopper" or "Cold Neutron" method [29 – 32]. The concept originated at Brookhaven, I think. A modified version was developed at Chalk River, largely by my colleague Alec T. Stewart. The method employs a polycrystalline filter of certain materials (notably beryllium metal) with particularly low neutron absorption and "parasitic" scattering, to eliminate all slow neutrons except those of such long wavelength (so slow) that they cannot be Bragg-scattered out of the beam by the filter. The scattered neutrons are then analyzed by a Chopper-based time-of-flight apparatus. In the Chalk River version two filters were employed (Be and Pb), in differential fashion.

An inverse of this method, the "Filter-Detector" or "Beryllium Detector" method [43] employs crystal diffraction for the monochromator and detects just "cold" neutrons. It has uses in studying those modes of motion in condensed matter of particularly high energy.

Absorption Methods

A number of elements exhibit large capture cross sections for slow neutrons. Boron and lithium and others have capture cross sections proportional to the wavelength; cadmium (and some rare earths) have neutron resonances with cross sections steeply varying with energy. When such absorbers intercept the neutrons scattered by a specimen, the measured intensity of the scattered neutrons depends on the neutron spectrum. In fact, the relative transmission of the scattered neutrons, for a sequence of absorbers running the gamut from very thin to very thick, is in essence related to the spectrum by a Laplace transform. The first experiments in neutron spectrometry, done in 1951 – 2 at A.E.R.E. Harwell by P.A. Egelstaff [21] and R.D. Lowde [23] and at Chalk River by Brockhouse and Hurst [22], employed such methods, but in very different ways. Only the last noted will here be discussed.

Over the range 0.2 eV to 0.5 eV the total cross section of Cd drops from over 6000 barns to less than 150. For an incident energy of 0.35 eV the transmission of Cd absorbers is quite sensitive to the scattered neutron spectrum. Fig. 2 shows transmission curves of annular Cd absorbers, for neutrons of initial energy 0.35 eV, scattered through a mean angle of 90 degrees, into an annular detector, by satisfactorily "thin" scatterers of lead, aluminium, graphite and diamond powder. The expected results for elastically scattered neutrons (with corrections for geometric effects) are also shown. (Some small

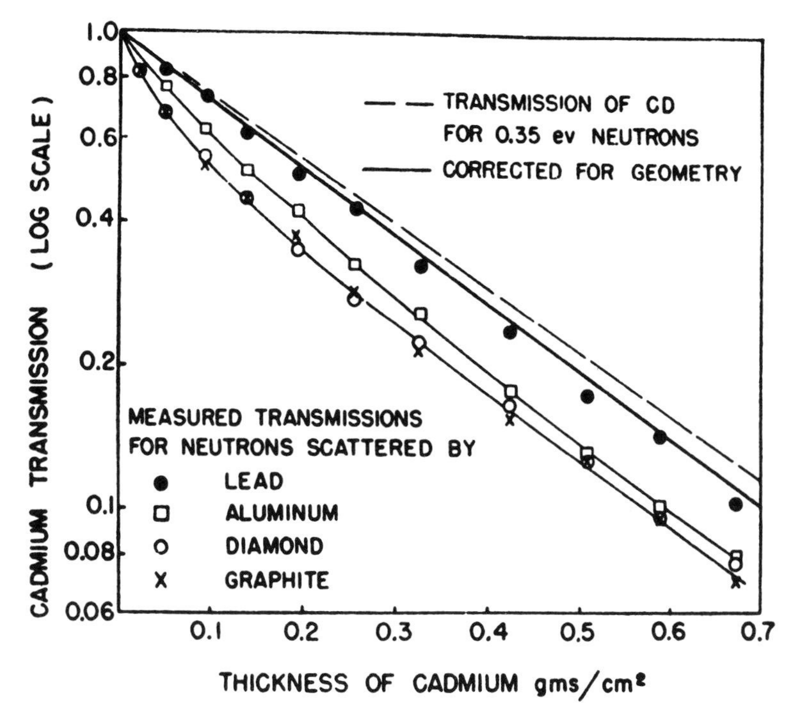

Fig. 2: The relative transmission of neutrons scattered into an annular detector by thin specimens of four materials, for annular Cd absorbers of various thicknesses, as well as the estimated transmission of elastically scattered neutrons. Each point is the result of measurement with a different thickness of absorber [22]. Scattering by the heavy element Pb shows little effect – as expected.

corrections for multiple event scattering were also made.) As expected, the neutrons scatter from the massive Pb atoms with little change in mean energy. For the lighter atoms Al and C, considerable energy transfer is indicated.

The results were compared with a calculation, by Finkelstein [5], of neutron scattering by a polycrystal, using the Einstein model for the atomic motions. In the Einstein model the atoms are considered to vibrate independently at a determined frequency and thus to be oscillators whose energies are integer multiples of an Einstein quantum energy (E_q), which can be estimated from data on the specific heat. In scattering the neutron can create or annihilate one or more quanta with consequent change in its energy. For Pb and Al multiple-quanta transitions predominate; for the two carbon specimens the scattering is mostly zero-quantum (elastic) with about 20% single-quantum. The scattered spectra so calculated can be used to generate Cd transmission curves.

For Pb the effects are small; the agreement is satisfactory but not very significant. For aluminium and the two forms of carbon, agreement was obtained with values of E_q which were compatible with specific heat data. At room temperature aluminium is in a "classical" regime ($E_q \ll kT$) and graphite and diamond are in a quantum regime ($E_q \simeq kT$), so the results here are very satisfactory.

Further calculations were made for the very unrealistic model in which the specimens were considered as ideal gases at room temperature, with atomic masses 12 for C, 27 for A1 and 207 for Pb. The continuous spectra obtained for Pb and A1 were very like the histogram spectra for the Einstein case, but the spectrum for the forms of carbon were markedly different. Thus the results suggested that details of the dynamical behaviour were not likely to be visible except where single-quantum scattering dominated.

Had available neutron fluxes been smaller there might well have developed a small field of absorption neutron spectroscopy, though it is unlikely that I would then have had the opportunity to give this lecture and to write this report. But events soon overtook the development as will now be discussed.

The Triple-Axis Spectrometer

As recounted earlier, the decision was made in 1951 to develop a neutron spectrometer which employed Bragg reflection by two single crystals as the monochromating units. The beams were to be two inches square, larger than with existing instruments. Where distances were short, the angular resolution was to be defined by Soller slits. The concept required low backgrounds and thus heavier shielding than was then employed. Monochromators would clearly have to be more efficient than the existing NaC1 ones, so a program for growing large metal crystals was instituted, which focused for the moment on lead and aluminium. Hurst was embarking on a low-temperature program, which looked forward to studies of the rare gas liquids, especially liquid helium. Chalk River was already well-supplied with liquid nitrogen; after some abortive attempts, a "Collins" Helium Liquefier was eventually installed. In 1952 an experienced low temperature physicist, David G. Henshaw arrived and took over major responsibilities for the crystal-growing and liquefier programs.

A primitive triple-axis spectrometer was set up at NRX in 1952. The intensities were too low for practical work, though the elastic peak from paraffin wax was seen and probably also that from vanadium. But in November 1952 the reactor suffered an accident and all experimental work terminated until the summer of 1954. Meanwhile Henshaw and Jack Freeborn had been successful in growing large single-crystal ingots of aluminium and lead. So that when work resumed at NRX both the powder diffractometer and the triple-axis instruments were empowered by much superior monochromators. Hurst and Henshaw embarked on studies of the rare gas liquids, instituting an ongoing program which, in successive forms and hands, continued for over 40 years.

The initial form taken by the spectrometer is shown in Fig. 3. The specimen is mounted on an indexed table which translates along the direction of the "monochromatic" neutron beam from the initial monochromator. Another single-axis spectrometer [7] which views the specimen through Soller slits, translates along rails parallel to the reactor face. The angle of scattering is calculated by triangulation and hence is changed only on occasion.

The triple-axis program met with almost immediate success. Some *ad hoc* additions to shielding were added, but otherwise the apparatus performed as expected. See Fig. 3. A variety of experiments were performed, some guided by existing theory and some not. A contributed paper, which turned into an invited paper [24], was given at the annual meeting of the American Physical Society in New York City in January, 1955. Preliminary results were presented for scattering from liquid and solid Pb, water, heavy water, vanadium and from certain paramagnetic materials with differing characteristics, including $MnSO_4$, Mn_2O_3 and MnO.

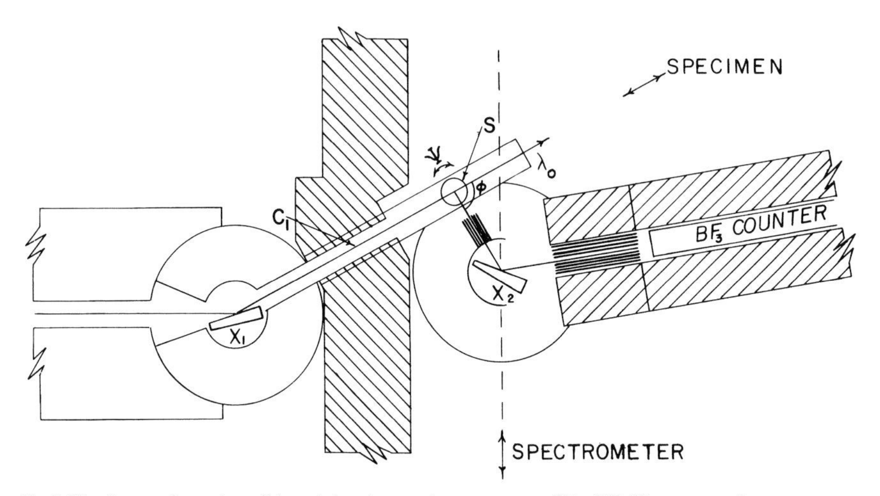

Fig. 3: The first crude version of the triple-axis crystal spectrometer [24 – 26]. Monoenergetic neutrons are selected by the large single crystal monochromator (X1) and impinge on the specimen (S), which is located on a table whose orientation (ψ) in the horizontal plane can be selected. This table can be moved along the direction of the incident beam as desired. The analyzing spectrometer, which employs crystal X_2, is a diffractometer (of especially large aperture) which can be translated as a unit; the angle (ϕ) through which the examined neutrons are scattered is determined by triangulation.

Several of these projects were continued and eventually reached publication, the first being the results for the two paramagnetics $MnSO_4$ and Mn_2O_3 [25]. $MnSO_4$ has a susceptibility fairly obedient to the Curie Law and hence it is expected that the interactions between the Mn^{++} ions are small; for the same reasons it was expected that the diffuse magnetic scattering should be almost elastic. In fact the scattering at angles smaller that the Debye-Scherrer lines had virtually the same energy width as the incoherant elastic scattering from vanadium. For Mn_2O_3 on the other hand there was substantial inelastic scattering with energy width consistent with a calculation of Van Vleck (15). Similar results were obtained for MnO, in which the energy-widths correlated with the short range order indications in the diffraction pattern. These latter results, continued successively with P.K. Iyengar and R.S. Weaver, did not get properly published, but were reported in review papers – perhaps constituting a kind of virtual publication?

From the beginning our program had been directed towards study of the frequency/wave number (energy/momentum) dispersion relation of the normal modes (lattice vibrations) in crystals, by means of study of the one-phonon scattering. (By analogy, it was assumed that spin waves in magnetic crystals could also be studied, as turned out to be the case.) In 1954 a theoretical paper by Placzek and Van Hove (9) discussed the phonon experiment in detail. More importantly, for us, they proposed a different kind of experiment, to determine the frequency distribution of the normal modes from the incoherent one-phonon scattering by a cubic crystal with one atom per unit cell. From a practical point of view, vanadium is the only natural element which is a candidate for this experiment, which was added immediately to our program.

Such an experiment was carried out [27] on the primitive triple-axis installation of Fig. 3, with the results shown in Fig. 4a. A strong elastic peak whose intensity decreases with increasing temperature, is visible – as expected from

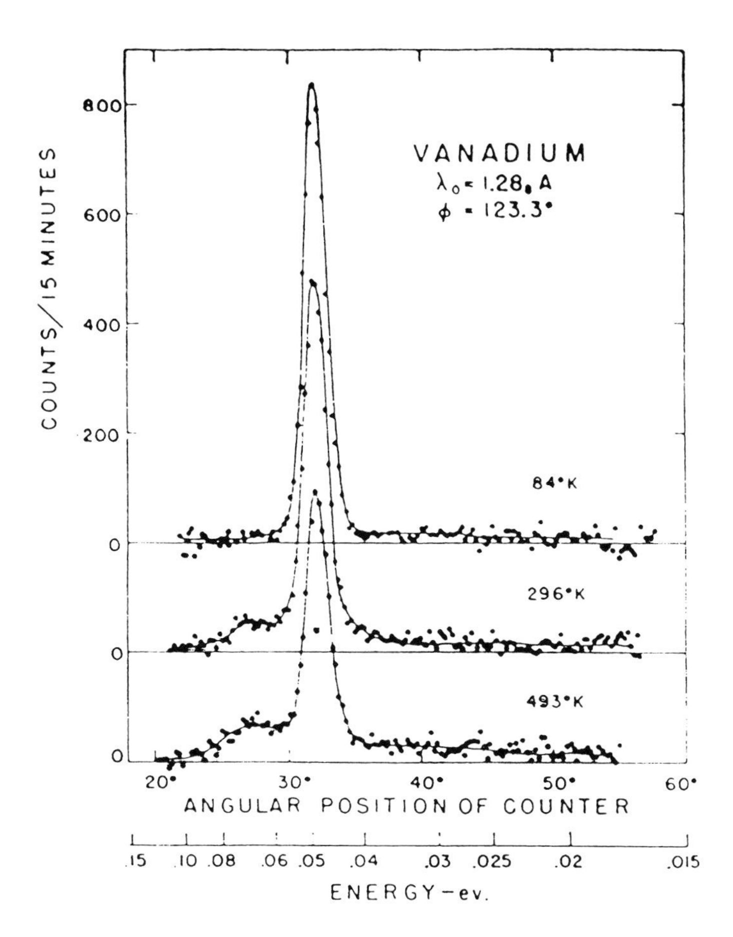

Fig. 4a: Measured energy distributions of neutrons scattered by vanadium as functions of the angular settings of the analyzing spectrometer, for three temperatures, with backgrounds substracted and intensities normalized [27]. The intensity of the elastic peak decreases with temperature in about the manner expected from Debye-Waller theory. The inelastic component is affected by the extreme dispersion of neutron spectrometers and requires transformation to a true energy distribution.

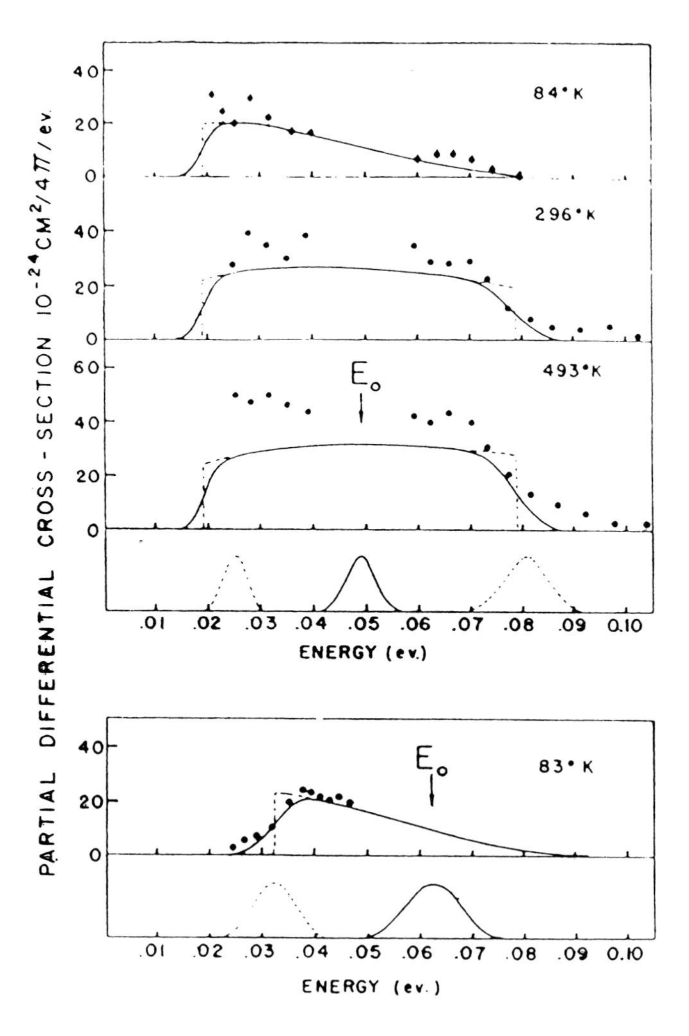

Fig. 4b: The transformed energy distributions (points) of the inelastic component in vanadium (Fig. 4). The curves are calculated (one-phonon) cross-sections for a Debye frequency distribution for the lattice vibrations. The elastic component provides a measure of the resolution function at the incident energy.

a Debye-Waller thermal cloud of increasing size. A weak inelastic component, whose intensity increases with increasing temperature, is also visible; it is this in which we are mainly interested. When this component is transformed from the distribution in (analyzer) angle to a distribution in energy, the curves of Fig. 4b are obtained. An Einstein model can be ruled out by inspection. One-phonon distributions calculated on a Debye model, with Debye temperature consistent with the Debye-Waller behaviour of the elastic peak, are also shown. (The additional scattering can reasonably be attributed to multi-phonon processes.) Thus the experiments supported the structure of the theory but the resolution did not permit improving upon the Debye model for the frequency distribution of the normal modes. Later experiments, using the filter-chopper method, by Stewart et al [32] at Chalk River and by Eisenhauer et al [31] at Brookhaven, showed plainly structure in the frequency distribution, indicating [see M – O] realistic structure for the lattice vibrations beyond the capabilities of a Debye model.

Least guided by theory and arguably the most important, were the experiments on liquids, particularly those on light and heavy water [33]. Here quasi-elastic scattering was found, whose energy widths could be qualitatively related to the manner in which diffusion went on in the liquid. Analogs to the Debye-Waller factor for the quasi-elastic component and the energy widths of the inelastic components in the scattering could be related to more general aspects of the thermal motions of the molecules. In the later work, by the author and by Sakamoto et al [33], a high-resolution rotating crystal spectrometer was employed, as well as the triple-axis instrument. The developments by Van Hove [13], of generalized (space-time vs momentum-energy) correlation functions, contributed to thinking about this notoriously difficult area, the physics of liquids.

The phonon experiment envisaged from the beginning – almost the raison d'etre of the program – was carried out on a large single crystal of aluminium, by Brockhouse and Stewart [28]. This material was selected for technical reasons: Al has little neutron capture and incoherent scattering and the coherent scattering cross section is small, so the physical size can be large without leading to large multiple scattering effects. And we had the crystal! The first results were obtained on the early crystal spectrometer with its fixed angle of scattering. The crystal was oriented with its (110) plane in the plane of the spectrometer and was turned a few degrees about an axis normal to that plane, to carry out successive experiments in successive orientations, as shown in Fig. 5. Neutron groups, with the expected properties of one-phonon

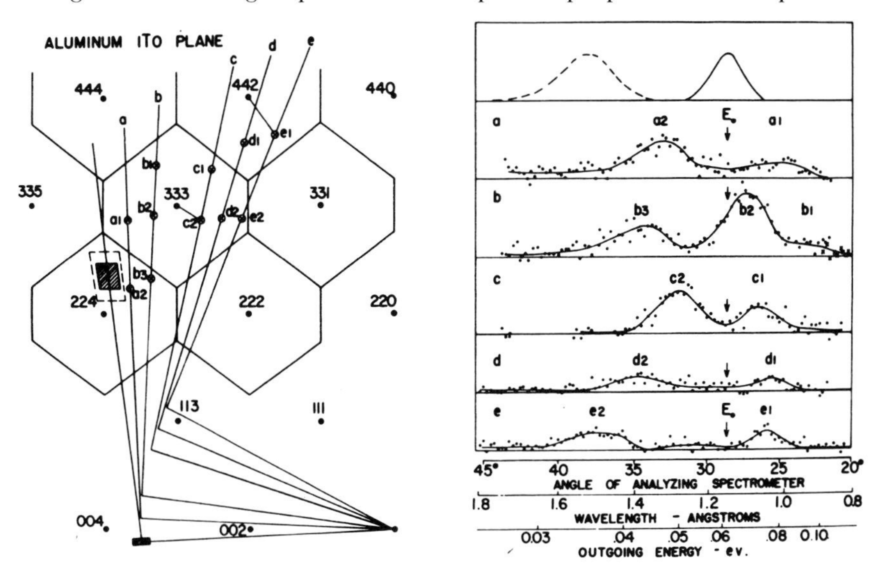

Fig. 5: Phonons in aluminium at room temperature [28]. (a) Some measured distributions in the (1,$\bar{1}$,0) plane of a single crystal at a fixed angle of scattering of 95.1 deg. (b) The reciprocal lattice for the (1,$\bar{1}$,0) plane with the crystal momenta plotted for the experiments in (a). One typical case is shown in detail as well as a crude depiction of the resolution in momentum.

groups, are observed in the different experiments, at different settings of the analyzing spectrometer and with different intensities. (Multi-phonon scattering would show a continuum.) From the momentum diagram the wave vector (q) of a phonon can be deduced and from energy conservation its frequency (ν). From a number of phonons with q in the same direction it can be made plausible (not shown) that the phonons obey a frequency/wave vector dispersion relation.

The experiments were continued on the differential filter-chopper as well as on the triple-axis apparatus, some 210 phonons distributed over two symmetry planes of the crystal's momentum space (reciprocal lattice). The results permitted refutation of the simple models for the interatomic forces then extant. The Born-Von Kármán theory itself can be taken to be phenomenological with innumerable possibilities for parameters, and thus hardly capable of refutation. If taken literally, as involving forces between ions in the crystal, then already these early results show surprisingly long-range behaviour for the interatomic force system. In the work intensities were used to assign polarization to phonons in symmetry directions. Henceforth measurements were made in symmetry directions as far as possible, in order to reduce the difficulties of analysis.

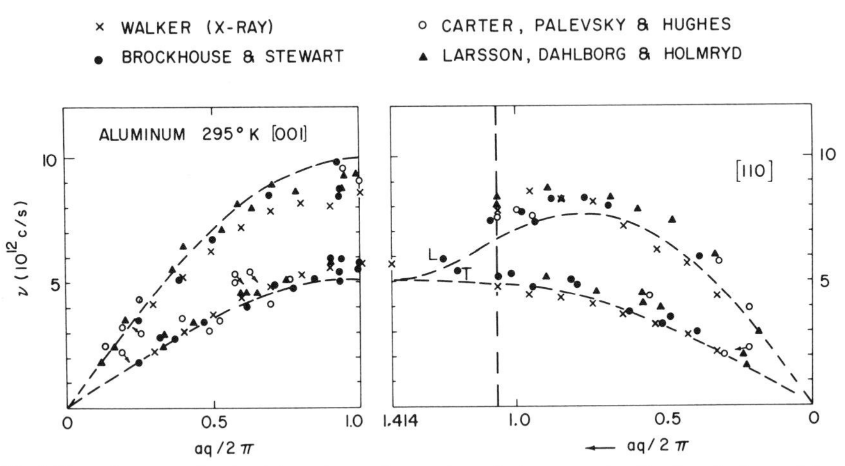

Fig. 6: Dispersion curves for aluminium at room temperature, for two symmetry directions in the crystal (collected results [28 – 30]. The neutron measurements from three laboratories are consistent with each other within the errors and are almost so with those inferred [18] from intensities of scattered X-Rays. Nevertheless the discrepancies would be very important in attempts to extract inter-atomic forces from the data.

Dispersion curves for the phonons in two symmetry directions of single crystalline aluminium are shown in Fig. 6, as determined in the above experiments and by Carter, Palevsky and Hughes at Brookhaven [29] and by Larsson, Dahlborg and Holmryd at Stockholm [30], both groups using the beryllium filter-chopper method. The various experiments agree within their still substantial uncertainties. Dispersion curves deduced by Walker [18] from his X-ray intensity measurements are also in substantial though not complete agreement. Because the frequencies from neutron experiments are deduced directly from the conservation rules, while those from X-rays require difficult calibrations and corrections, the neutron frequencies are believed to be the more reliable in principle.

Experiments were carried out to study spin waves ("magnons") in single-crystal magnetite (Fe_3O_4). This cubic ferrite ("lodestone") was selected because large single crystals of natural origin were available. It was expected that a more or less isotropic, quadratic dispersion curve would be found at small wave vectors, close to positions in reciprocal space where strong Bragg scattering occurred. This was indeed found [39] to be the case. In a later experiment [39] it was verified that the neutron groups concerned showed the change of intensity expected when a suitable magnetic field was applied and thus that the quanta concerned were indeed magnons and not phonons.

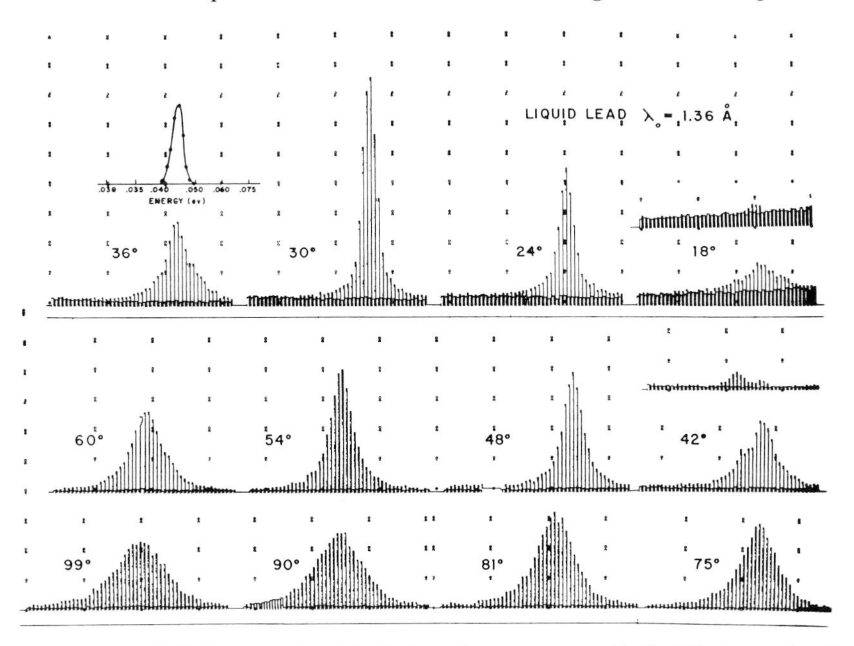

Fig. 7: Histograms [43] of energy-analyzed distributions of neutrons scattered by liquid lead at a series of scattering angles [34]. The scattering angles are roughly related to the momentum transfer Q, the abscissa angles to the energy transfer. Appropriate measured backgrounds are also given. The resolution function measured with a vanadium scatterer is shown at the upper left.

Early results for liquid lead were not easily analyzable. Because of the large mass of the Pb atom, the energy transfers are small. The narrow inelastic component would not be separable from a quasi-elastic component, if the latter existed, because of the instrumental energy resolution. So the early results were never published, though they were presented at Conferences. But with much improved facilities available, new measurements (Fig. 7) with Noel K. Pope [34] were analyzed using the Van Hove [13] transformations to give directly the self and pair correlation functions of the atoms in liquid lead (Fig. 8). The temporal development of these functions, from the defined origin at t=0 to the vanishing of correlation some ten pico-seconds later, provides a sort of statistical "movie" of the atomic motions in the liquid.

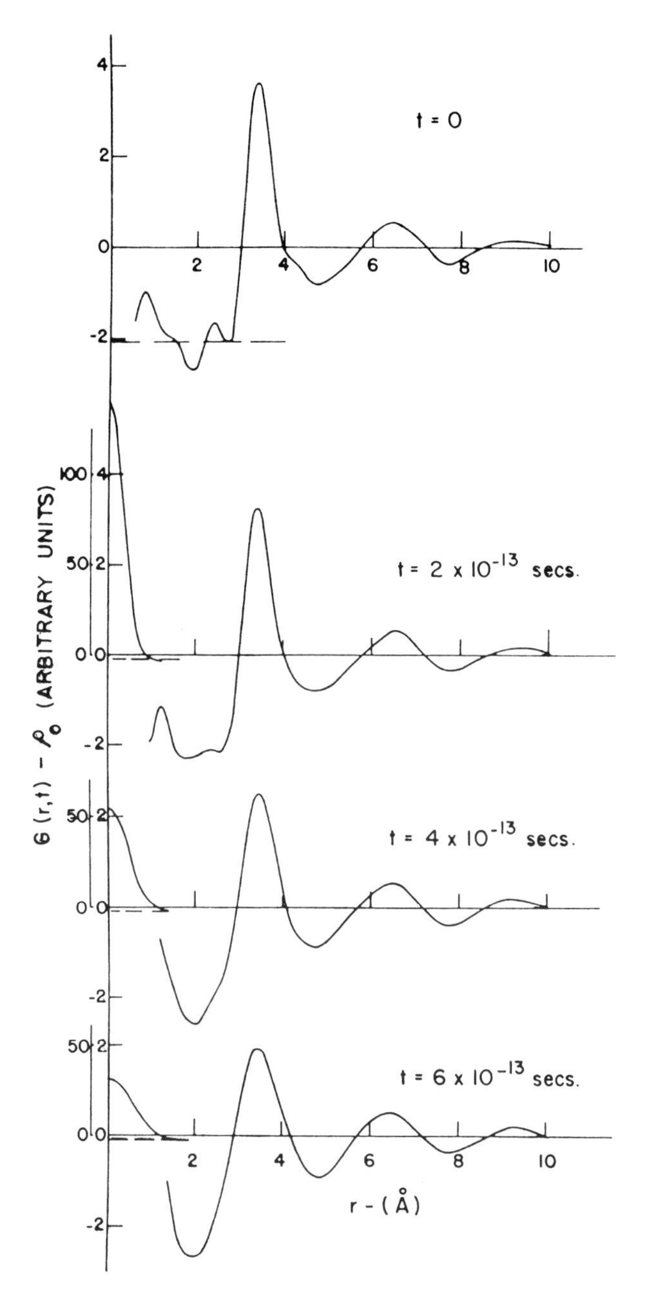

Fig. 8: Van Hove transformations of the data of Fig. 8 [34]. The distribution for time t=0 is essentially the familiar pair distribution of X-Ray scattering experiments. The peak (δ-function) at the origin gradually diffuses outwards with elapsed time t as do the first and subsequent neighbour peaks in a sort of moving picture of what goes on in the liquid.

In a wide-ranging memoir such as this, it would not be right to omit mention of the work [35 – 38] on liquid helium, even though I had little part in it. From early on, Henshaw and Hurst had looked forward to including study of the inelastic scattering of neutrons by liquid He as part of their program on the rare gas liquids. But the experiment looked formidable, both as to execution and analysis of results. This appraisal changed abruptly with the publication of a theoretical paper by Cohen and Feynman [35], who predicted a "line" spectrum instead of the continuous spectrum of unknown behaviour which had been assumed. Preparatory work was immediately started. Palevsky, Otnes and Larsson [36], at Stockholm, found the predicted line ("roton") spectrum, using the Beryllium-Chopper method. This was followed closely by work by Yarnell et al [37] and by Henshaw [38], improving and extending the Stockholm results. At Chalk River work has continued on various aspects of liquid helium physics alsmost to this day.

Developments in Apparatus and Methods

In 1956 the instrument had been completed by construction (by the Engineering Design Branch at Chalk River), of a proper analyzing spectrometer which was set upon the arm of the original [7] single-axis instrument, whose "half-angling" table now carried the specimen. With this new triple-axis instrument, all variables were motor-driven except the initial monochromator on the first (fixed) axis. Thus the initial energy was changable only by changing the monochromator. When the spectrometer was moved to the NRU reactor in 1959 the first axis was also engineered (by W. McAlpin) so that the initial energy (E_0) was also controllable within an experiment. Fig. 9 shows the spectrometer in place at hole C5 of the NRU reactor. Fig. 10 shows a schematic diagram of the arrangement. It might be noted that the relative arrangement of the incident and outgoing beams has importance because of correlations which affect the resolution in momentum space.

In all the work the intensity of the incoming neutron beam is monitored by means of a "thin" neutron detector of known efficiency. In the experiments so far discussed, it was the analyzing spectrometer which was changed and consequently the outgoing energy (E') which was varied. Corrections to the intensities for the efficiency of the analyzer had to be made. For a crystal spectrometer these are not readily calculable, so suitable auxiliary measurements were often required. If the analyzer is kept fixed, its efficiency does not enter the problem unless absolute partial differential cross-sections are desired. So from 1959 on distributions were often measured at constant outgoing energy.

It had been recognized in crystallography that wave-vector (momentum) transfers, not the angles of scattering and crystal orientation, were the "natural variables". This had shown up also in our original consideration, in 1950 – 51, of the theoretical analysis by Weinstock [19] and others, of the scattering by phonons in crystals. Over the years it came to seem obvious that the

Fig. 9: Photograph (1959) of the original triple-axis spectrometer at the NRU reactor at Chalk River. The bank of 52 rotary switches, preset to go through an energy scan of up to 26 points, can be seen in the upper centre. One of the three variables involved would traverse linearly through the domain of settings desired, while the other two advance nonlinearly according to the settings of the appropriate switches. (AECL photo)

neutron, in scattering, conserved energy and momentum with the specimen scatterer – the transferred momentum (Q) might be carried by the specimen as a whole or by entities within it or by a combination of the two. The specimen being so large, the energy transferred ($\omega = E_0 - E'$) is inevitably carried by the quasi-particle entities within it. When no such entities are involved, the scattering is "elastic". So in thinking about an experiment, one can start with energy-momentum (ω, Q) conservation, not have to reach it at the end of a calculation. This point of view was re-enforced by the evocative Van Hove transformations. The intensities are then governed by a generalized structure factor analogous to the structure factors of neutron and X-ray crystallography. Of course, as for quasi-elastic scattering, some care must be used, in such a semi-classical approach, that the necessary internal entities are considered.

This was the era of the early development of digital computers. Chalk River obtained such a facility about 1954; programs written in a predecessor of Fortran could be run on the Datatron to print out "hard copy" results of a sequence of calculations. The calculations could provide a sequence of angular increments of the variables of a triple-axis spectrometer: the angles of the analyzer and the initial monochromator, the angle of scattering and the angular setting of the specimen in the plane of the spectrometer. Fig. 10 shows the schematic diagram.

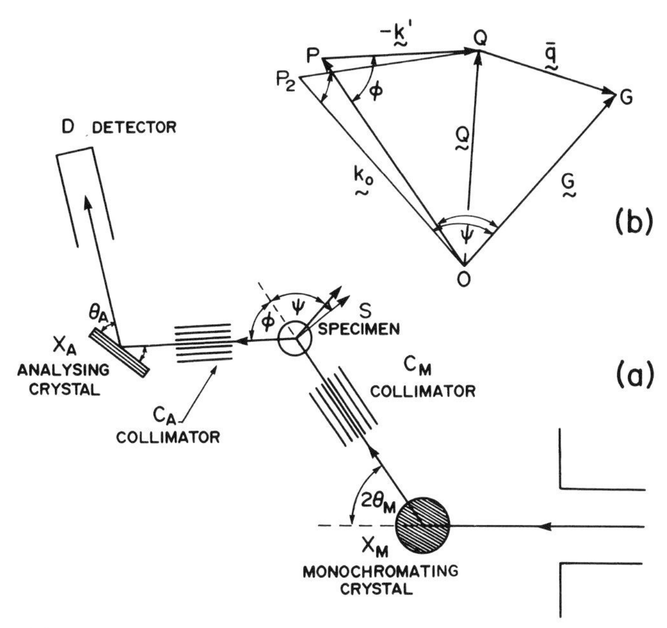

Fig. 10: Schematic drawing of a triple-axis spectrometer together with the momentum space diagram in the same co-ordinate system. The dashed lines indicate a second point on a "Constant Q" curve at fixed incoming energy [43]. The massive shielding required, particularly around the monochromator, is not shown.

In operation one variable would be held constant, one variable moved linearly in constant increments; the other two variables would move non-linearly in the increments provided by the computer. Initially these were set by means of 52 rotary switches which can be seen in the upper centre of Fig. 9. Thus an experiment ("scan") could involve up to 26 points. A stepping switch would accept the two instructions in sequence and match them with the number of microswitch closings by the angular variables concerned. The overall angular traverse was set by the number of increments traversed by the angular variable using fixed increments. This system, largely designed by E.A. Glaser, was brought into operation in 1959. A bit later a system designed by W.D. Howell, which employed punch tape put out by the computer to transport the data to the spectrometer, was in turn installed. Fig. 10 shows the initial arrangement for a scan in which the analyzer setting would be changed linearly and the angular variables set to give the desired momentum transfer. A second setting is indicated for a second point on the scan. And so on.

The method can be thought of in optical analogue. There is an incoming beam of "green" neutrons and the analyzer goes through a "rainbow" sequence, the angular variables meanwhile being changed sequentially to keep the momentum transfer Q constant. Alternatively the analyzer is set to detect "green" neutrons and the incident neutrons go through the "rainbow" sequence. In the simple theory the results are equivalent, but technical matters may favour one over the other.

The variables momentum transfer Q and energy transfer ω are often given in units of Planck's constant ħ and are then referred to as wave vector and (angular) frequency respectively. The frequency ν is then equal to ω/2π

Subject to some geometrical limitations, an experiment could in principle be carried out over any track in Q-ω space. But the most useful has been the "Constant-Q" or constant momentum transfer method. (The method of constant energy transfer has also been used, especially for steep dispersion curves.) The major advantage of these methods [43] is, of course, that the experimenter can get the data wanted for the analysis proposed, rather than having the experiment turn up what might be a great deal of unwanted data. Of course, also, there is a downside to this – something unexpected might be missed. Good practice thus involves giving in any report the value of the fixed variable, usually the incident or outgoing energy, even though it does not appear explicitly in the analysis. (It may affect contaminant scattering.) There is another important virtue of the Constant-Q method: for important technical reasons, the accuracy is greater and the intensities are more easily interpretable than with ordinary neutron spectrometry. (These facts are connected with the occurrence of the Waller-Fröman Jacobian [17] in the theory.)

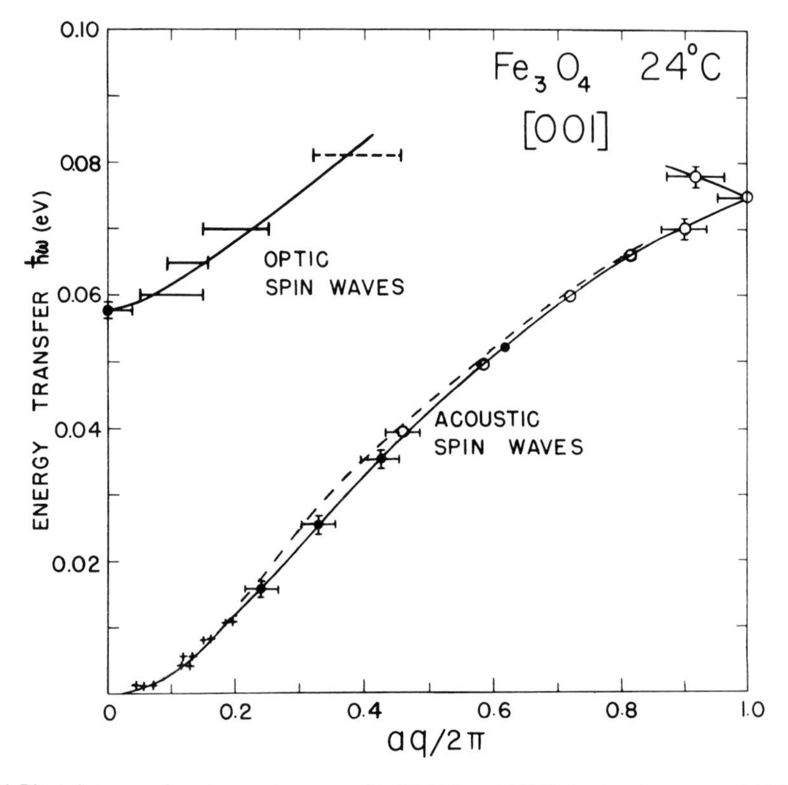

Fig 11: Dispersion curves for spin waves in a magnetite (Fe_3O_4) crystal [40], showing the expected initial quadratic behaviour and the initial part of an optical mode.

With the advent of the Constant-Q ability and the considerable improvements brought about by the move to the new NRU reactor, experiments with much improved resolution and precision were possible. Technical advances were now concerned with making more precise certain "rules of thumb" regarding selection of points in momentum space for optimum resolution, discrimination between different symmetry types of neutron groups, and so on. Of course, as new components became available, such as ^{3}He proportional counters and better monochromators, these were eagerly adopted.

Using the new methods and the improved facilities at NRU, Watanabe and Brockhouse [40] were able to extend considerably the earlier work [39] on the spin waves in magnetite. See Fig. 11. The complete "acoustic" dispersion curve was obtained as well as part of an "optical" branch near the zone centre, both in qualitative agreement with theoretical expectations [4, 40].

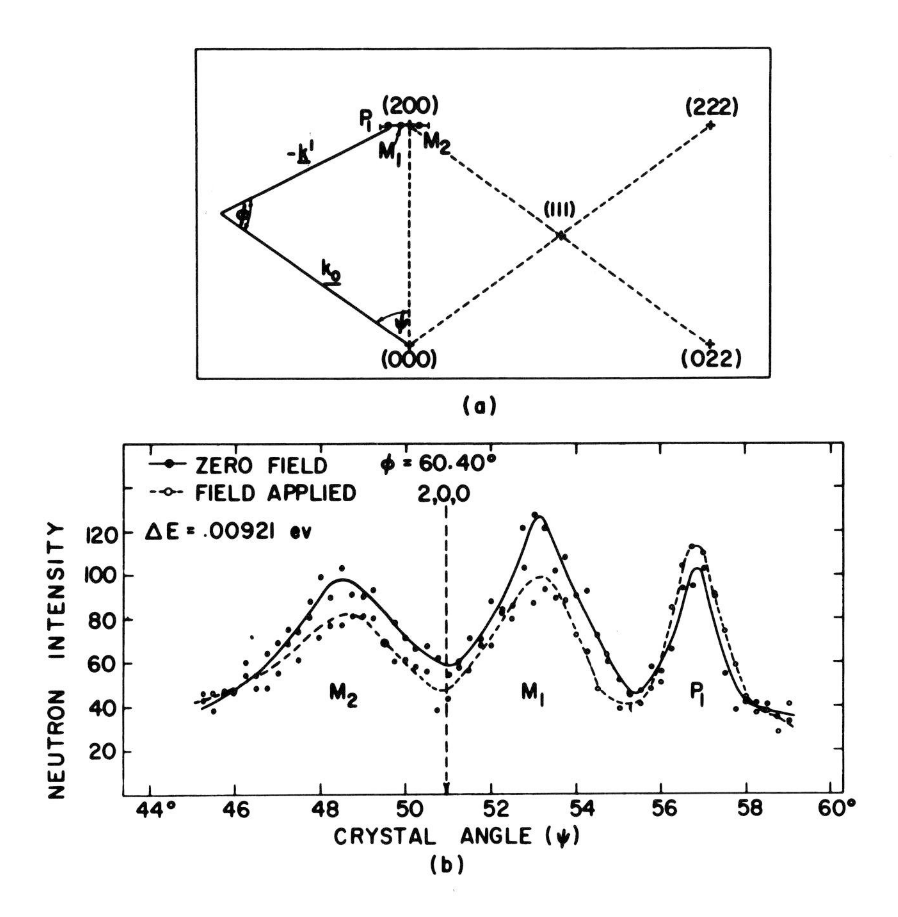

Fig. 12: Momentum diagram (a) and neutron groups (b) for a crystal of (F.C.C.) Co (8% Fe). The magnetic signatures on the intensities indicate that M_1 and M_2 are groups for spin wave quanta (magnons) and that P_1 is a phonon group [41].

At the time it was still an open question whether spin waves existed in metals. Sinclair and Brockhouse [41] studied the neutron scattering in a FCC cobalt-iron alloy (8% Fe). (This crystal was available for use as a neutron polarizer. The magnetic and nuclear scattering are comparable and this fact over-rode the considerable objection that Co has a large capture cross section for neutrons.) The rather steep spin wave dispersion curve could be measured only at small wave vectors (to about 20% of the zone centre). See Fig. 12. It was established from the "signature" under an applied magnetic field that the neutron groups indeed represented spin wave quanta (magnons).

Experiments of other types were carried out. The crystal field spectra of a number of rare earth oxides were studied [42], but analysis proved intractable and these will not be discussed here. Watanabe and Brockhouse [40] studied the exchange field splitting in Ytterbium Iron Garnet. And a visiting colleague, L. N. Becka studied rotational modes in a number of organic compounds. But the major target of my program from here on was phonon physics.

Phonons, Dispersion Curves and Interatomic Forces

Already, before the major acquisitions of 1959, there had been phonon studies which, though desirous of improvement in resolution and precision, were fully competent to provide important instruction on the physics of the specimens concerned. The first of these was a study of the lattice vibrations in two symmetry directions of germanium, by Brockhouse and Iyengar [45]. See Fig. 13. The dispersion curves for the acoustic and optical branches were

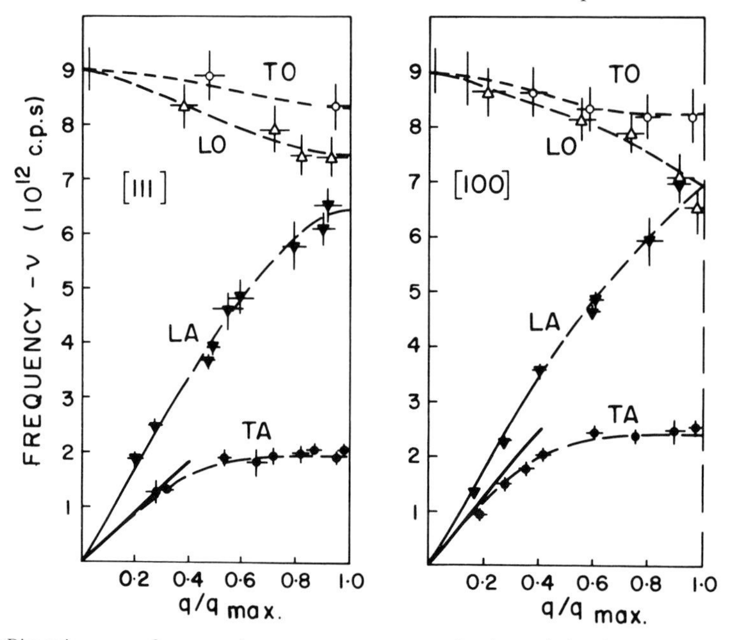

Fig 13: Dispersion curves for germanium at room temperature, showing optical and acoustic branches in two symmetry directions [45].

well delineated and were analyzed in terms of simple interatomic force constant models from the literature, but without complete success, indicating some missing factors in the models. Some elegant connections with far infrared spectra were brought out and the accepted placements of the maximum of the valence band and the minimum in the conduction band were supported. Further experiments were concerned with the phonons in silicon [46], and the homology found to exist with those in germanium, and with the temperature dependence in the phonons in germanium [47]. It should be noted that some results from Brookhaven (using the Filter-Chopper method) are noted in [45].

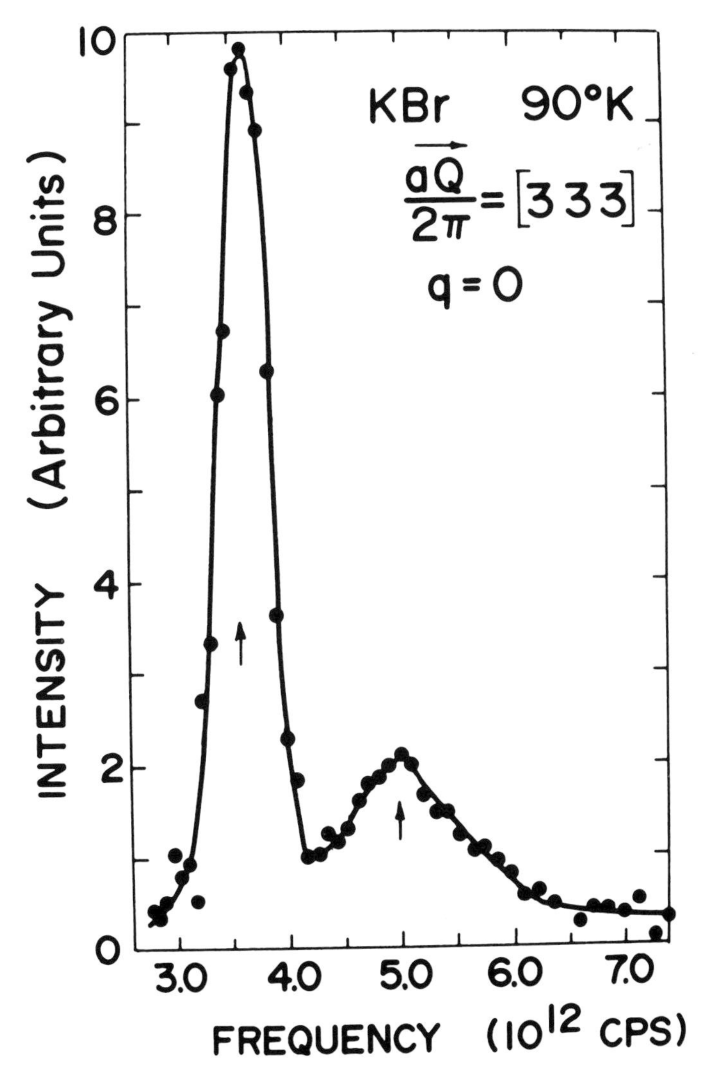

Fig. 14: The zone-centre phonons in a crystal of KBr [50]. The transverse optical mode is active in Infrared spectroscopy but the longitudinal optical mode is not. Both are visible in this neutron pattern and the ratio of their two frequencies is about as predicted from theory [O].

An extensive program was undertaken on alkali halide crystals. In selecting which of these to study, attention was paid to the neutron properties and to the mass ratio between the two component atoms. Two crystals were studied extensively, sodium iodide (NaI) initially [49] and then potassium bromide (KBr) [50]. In the course of this work our visiting colleague, William Cochran invented his famous and useful "Shell Model" for polarizable ions in crystals. One of the results of these studies was the verification of the well-known Lyddane-Sachs-Teller formula [O] for the zone centre optical phonons in terms of the dielectric constants. See Fig. 14.

An extensive program on metals was carried out, with special emphasis on the metal (sodium) then considered to be the "simpliest" and on a metal (lead) in which the phonon-electron interaction was known to be appreciable (from the fact that it is a superconductor with high transition temperature). And, as always, an important consideration was the availability of suitable single crystal specimens. We had lead crystals from Henshaw's work on monochromators and Raymond Bowers of Cornell University was able to grow the large crystals of sodium required.

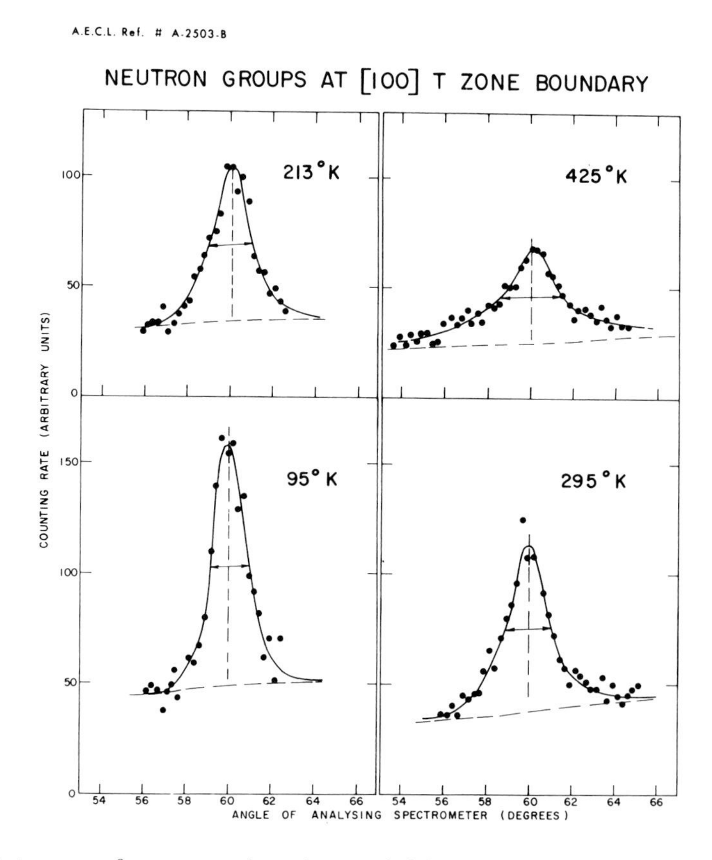

Fig. 15: Neutron groups for a symmetry phonon in a crystal of Pb at four temperatures showing line-widths increasing with temperature [60]. This receives a natural interpretation as a reduction with temperature of the lifetime and coherence length of the phonon involved, because of interaction with other phonons.

Measurements were made on Pb [51] and Na [54], at a low (liquid nitrogen cooled) temperature for all the symmetry directions available including some on the surface of the Brillouin zone. In both cases measurements were made also at higher temperatures, to study thermal broadening and shifts in the phonons. See Fig. 15. Substantial effects were observed, which proved difficult to interpret quantitatively in terms of atomic forces. For both metals at high temperatures, the phonon lifetimes and mean free paths became remarkably short at wave vectors far from the zone centre. Similar results for Al were obtained independently by Larsson et al [30], using the Filter-Chopper method. Our high temperature results received only incomplete or late publication.

Analysis of the results for Pb [51] displayed an interatomic force system of long range and great complexity. Calculations for alkali metals by Walter Kohn [52] suggested that anomalies in the phonon dispersion curves, arising from electron-phonon interaction, might be found at wave vectors related to the Fermi surface of the metal. A series of measurements were made to search for such anomalies in the dispersion curves of Pb. The Kohn anomalies were indeed found [53] in repeated experiments at more or less the expected phonon wave vectors. See Fig. 16. A search was made [54] also in Na (for which the electron-phonon is believed small) without positive result, supporting to some degree the assignment for Pb.

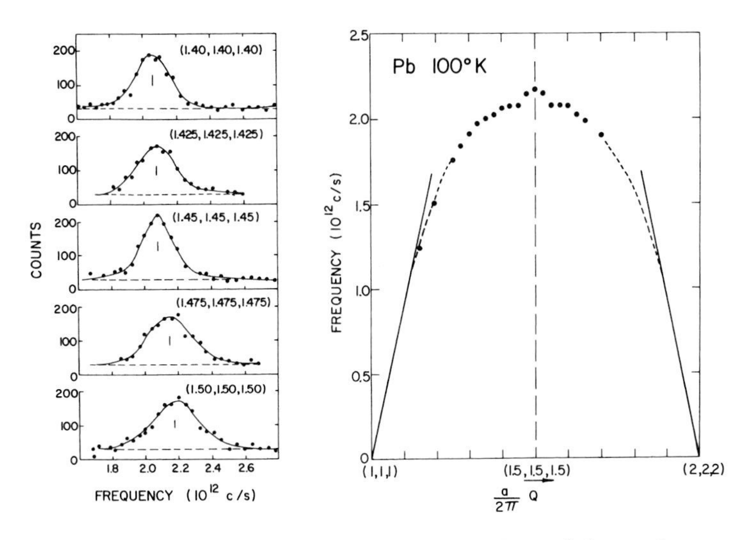

Fig. 16: Kohn Effect in dispersion curves of Pb [53]. (a) A series of closely-spaced phonons show anomalous variation (b) in the corresponding dispersion relation ν(q), in this case at a calliper dimension of the Fermi surface in the [ζ,ζ,ζ] direction.

The results of Woods et al [54] for sodium could be analyzed in terms of a Born-von Kármán model of much shorter range than for lead. Dixon et al [55] used the model to calculate the frequency distribution of the normal modes, which gave excellent agreement with specific heat results from the literature. See Fig. 17. The computer calculation had a resolution capable of showing the topologically significant Van Hove critical points [14] with precision.

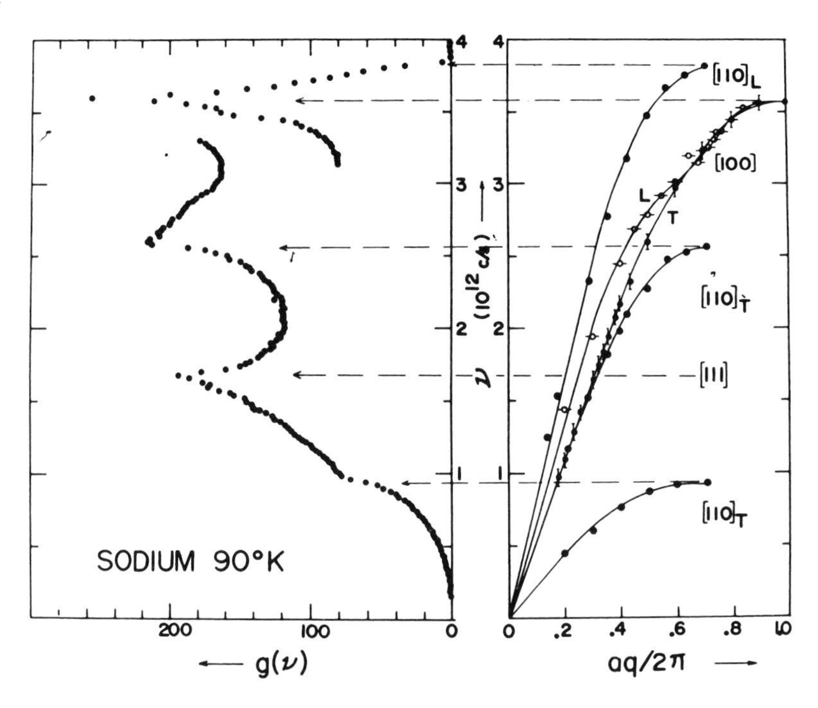

Fig. 17: The frequency distribution g(ν) for Na at 90 K [55], calculated from a force constant model fitting the dispersion curves ν(q) also shown [54]. The relation between the frequencies at symmetry points and the Van Hove [14] singularities in g(ν) are exhibited.

Finally Fig. 18 shows a comparison of the complete results for (body-centred cubic) Na and those obtained, on the same apparatus by Gilat and Dolling [56], for β-brass (CuZn) ordered with copper mainly on the cube corners and zinc on the body centres. It will be seen how much of the general appearance comes about from symmetry; the effects of what must be considerably different interatomic force systems is for the most part hidden in the fine details. (This is not always the case – Na and CuZn are both (probably) nearly-free electron metals. Drastically different binding schemes can produce much greater effects.) The curves of Fig. 18 point up the effects of the symmetry breaking by the difference introduced by the ordering of the Cu and Zn atoms on the crystographically equivalent cube corners and body centres.

The results of Figs. 13 – 18 illustrate a basic fact about neutron spectroscopy of crystals. Despite the considerable amount of data shown, it should be realized that vastly more is potentially available in the off-symmetry frequencies and in the intensities of the neutron groups. But presentation and analysis are alike difficult: Presentation, because our intuition and visualization

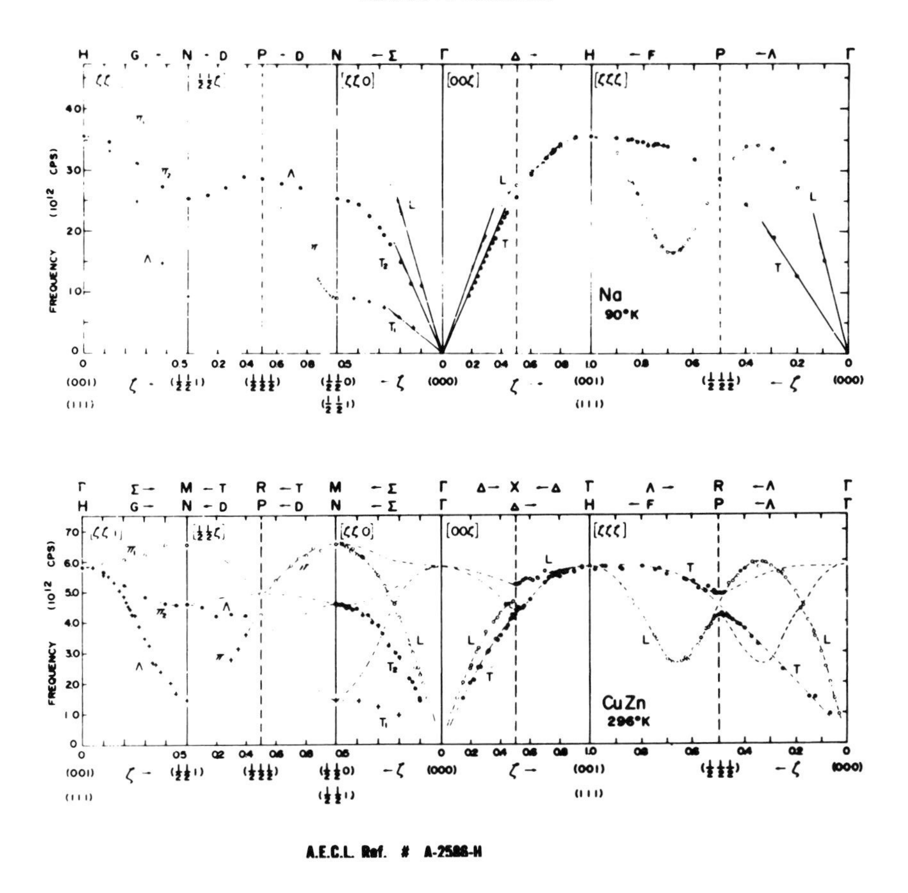

Fig. 18: Complete dispersion curves for the symmetry directions in Na [54] and in ordered CuZn [56]. The group-theoretical notation for the various branches are given as letters. Dashed lines without points represent measurements in symmetry-related directions of the zone.

processes are limited to essentially three spatial diemensions and counted time and the data for crystals with one atom per cell require five dimensional presentation and, for more complicated crystals, still more. Analysis, because it is not possible generally to assign a particular frequency or intensity to a particular formula – there is sorting-out to be done first. Several of the works cited go into this in a small way, via calculations and experimental determinations of dynamic structure factors.

These works and others done over the next few years, were extensively reported in review articles [60 – 63] In these years also comparable results began to appear [48, 57] from other laboratories. The field was now a field in being and not merely in development. My own program, now centred at McMaster University, continued to focus on phonon physics with emphasis on metals and alloys. Reference is made to work [44] by two of my first students there, Eric Svensson and Michael Rowe, on phonons in crystals with point mass defects. The initial experiments were on single crystals of an alloy of copper with 9.3 Atomic % of gold. The fraction of defects was larger than might be wished but interesting results were obtained, which Svensson and Rowe subsequently improved upon.

Nature, Neutron Spectroscopy and the Grand Atlas

The result of human action is the reward for that action. In physics experimentation we are guided in our choice of action by the existing theory and by calculations based upon that theory. If things turn out as expected the reward is additional confidence in that theory. If the result is not as expected then new possibilities for experimentation and theory-construction are opened up. The early work on neutron physics resulted in little to challenge the basic theories of physics (quantum mechanics in particular) but did, I think, add somewhat to the confidence in which that theory was held. With the models to be found in the literature of the time, the situation was different. I cannot recall a single instance in which experimental results were in quantitative agreement with a pre-existing model calculation. Indeed, the results were usually qualitatively different, but in a manner in which the differences could indicate the ways in which the model needed modification if it were to describe the experimental results. The existing theory normally functioned well as a language in which to discuss the experiments, but had to be extended if it were to describe the results satisfactorily. And only for phenomenological theory, with the possibility of numerous fittable parameters (as the Born-von Kármán theory), would the extension be "trivial".

The fact that theory functioned satisfactorily as a language meant that sometimes there was a possibility to compare directly neutron results with those from other types of experiment (such as infrared absorption) and thus, in a sense, to give an interpretation of the latter. But the major importance of the neutron experiments derives from their influence on the introduction of new microscopic models which can be applied to calculations of other quantities as well. With the vast numbers of "universes of discourse" (physical systems at a particular temperature, under a particular applied pressure and son on) the question arises: how to decide what universes of discourse should be included in the Grand Atlas and to what extent. Chemical and crystallographic structure seem already to have been admitted for any and all such "universes". Are phonons and their dispersion curves and decay descriptions (their "laws"), magnons and their "laws", the numerous other quantum entities and their "laws" to be admitted? Are time-dependent distribution functions also to be admitted? Or are we to act as if the basic theories fully contain all these and that only those quantities are to be considered for inclusion which go to characterize the system? I have had opinions on these matters but that is another story.

ACKNOWLEDGMENTS

In addition to the numerous colleagues who contributed long ago to the work discussed, many of whom are acknowledged in the text and references, I should thank also Atomic energy of Canada Ltd. and McMaster University and several of their members, for help in meeting the many demands arising from this remarkable event. In particular, Mrs. Pat Carter gave assistance with this manuscript and other secretarial necessities. Dr. William Buyers and Dr. Gerald Dolling were helpful in many ways, including provision of quality copies of several figures, especially Figs. 1 and 9, the two photographs.

A: The Neutron and its Applications, 1982, editor P. Schofield. The Institute of Physics, Bristol and London (Conference Series Number 64), 1983. Also: Pile Neutron Research in Physics. International Atomic Energy Agency, Vienna, 1962.
B: Fifty Years of Neutron Diffraction, editor G.E. Bacon. Adam Hilger, Bristol, 1986.
C: Inelastic Scattering of Neutrons in Solids and Liquids. International Atomic Energy Agency, Vienna, 1961.
D: Inelastic Scattering of Neutrons in Solids and Liquids. International Atomic Energy Agency, Vienna, Vols. I,II: 1963, I,II: 1965; I,II: 1968, 1972.
E: Proc. of International Conference on Magnetism and Crystallography Vol. II, III: 1961. (Jour. Phys. Soc. of Japan 17: Suppl. B-II, B-III, 1962.)
F: The Interaction of Radiation with Solids. Editors R. Strumane, J. Nihoul, R. Givers and S. Amelinckx. North-Holland Publishing, 1964.
G: Neutron Physics, editor M.L. Yeater. Academic Press, New York, 1962.
H: Solid State Physics, editors F. Seitz and D. Turnbull. Academic Press, London and New York, 1955 – 1994.
I: Phonons and Phonon Interactions, editor Thor A. Bak. W.A. Benjamin Inc. New York and Amsterdam, 1964.
J: Phonons in Perfect Lattices and in Lattices with Point Imperfections, editor R.W.H. Stevenson. Oliver and Boyd, Edinburgh and London, 1966.
K: Thermal Neutron Scattering, editor P.A. Egelstaff. Academic Press, London and New York, 1965.
L: Magnetic and Inelastic Scattering of Neutrons by Metals, editors T.J. Rowland and P.A. Beck. (Metallurgical Society Conferences, Vol. 43). Gordon and Breach, New York, 1968.

-MONOGRAPHS AND TEXTS –

M: Properties of Metals and Alloys. N.F. Mott and H. Jones. Oxford University Press, 1936.
N: The Modern Theory of Solids. F. Seitz. McGraw Hill, New York and London, 1940.
O: Dynamical Theory of Crystal Lattices. Max Born and Kun Huang. Oxford University Press, 1954.
P: Neutron Diffraction. G.E. Bacon. Oxford University Press. Editions: 1955, 1962, 1975.
Q: Scattering of Thermal Neutrons (A Bibliography 1932 – 1974), editors A. Larose and J. Vanderwal. Solid State Literature Guides Volume 7, IFI/PLENUM, New York and London. 1974.

REFERENCES

1. Brockhouse B.N. Slow neutron spectroscopy: an historical account over the years 1950 – 1977. Source A:193-198, 1982.

– PRELUDE OR THEORY –

2. Cassels J.M. The Scattering of Neutrons by Crystals. Prog. Nucl. Phys., editor O.R. Frisch. 1:185-215, 1950.
3. De Gennes P.G. Liquid Dynamics and Inelastic Scattering of Neutrons. Physica 25:825-839, 1959.
4. Elliott R.J. and Lowde R.D. The inelastic scattering of neutrons by magnetic spin waves. Proc. Roy. Soc. (London) A230:46-73, 1955.
5. Finkelstein R.J. Scattering of Neutrons in Polycrystals. Phys. Rev. 72:907-913, 1947.
6. Halpern O. and Johnson M.H. On the Magnetic Scattering of Neutrons. Phys. Rev. 55:898 – 923, 1938.
7. Hurst D.G., Pressesky A.J. and Tunnicliffe P.R. The Chalk River Single-Crystal Neutron Spectrometer. Rev. Sci. Instr. 21:705 – 712, 1950.
8. Moorhouse R.G. Slow Neutron Scattering by Ferromagnetic Crystals. Proc. Phys. Soc. A64:1097 – 1107, 1951.
9. Placzek G. and Van Hove L. Crystal Dynamics and Inelastic Scattering of Neutrons. Phys. Rev. 93:1207 – 1214, 1954.
10. Seegar R.J. and Teller E. On the Inelastic Scattering of Neutrons by Crystal Lattices. Phys. Rev. 62:37 – 40, 1942.
11. Shull C.G. Wave properties of the neutron. Source A: 157 – 168, 1982. Early Neutron Diffraction Technology. Source B: 19 – 25, 1986.
12. Squires G.L. Multi-oscillator processes in the scattering of neutrons by crystals. Proc. Roy. Soc. (London) A212:192 – 206, 1952.
13. Van Hove L. Correlations in Space and Time and Born Approximation Scattering in Systems of Interacting Particles. Phys. Rev. 95:249 – 262, 1954. Time – Dependent Correlations between Spins and Neutron Scattering in Ferromagnetic Crystals. Phys. Rev. 95:1374 – 1384, 1954.
14. Van Hove L. The Occurence of Singularities in the Elastic Frequency Distribution of a Crystal. Phys. Rev. 89:1189 – 1193, 1953.
15. Van Vleck J.H. On the Theory of the Forward Scattering of Neutrons by Paramagnetic Media. Phys. Rev. 55:924 – 930, 1939.
16. Vineyard G.H. Scattering of Slow Neutrons by a Liquid. Phys. Rev. 110:999 – 1010, 1958.
17. Waller I. and Fröman P.O. On neutron diffraction phenomena according to the kinematical theory. I. Arkiv for Fysik 4:183 – 189, 1952. Fröman P.O. ibid II. 4:191 – 202, 1951.
18. Walker C.B. X-Ray Study of Lattice Vibrations in Aluminum. Phys. Rev. 103:547 – 557, 1956
19. Weinstock R. Inelastic Scattering of Slow Neutrons. Phys. Rev. 65:1 – 20, 1944.

– GENESIS –

20. Brockhouse B.N. A Childhood of Slow Neutron Spectroscopy. Source B:35 – 46, 1986.
21. Egelstaff P.A. Inelastic Scattering of Cold Neutrons. Nature 168:290, 1951.
22. Brockhouse B.N. and Hurst D.G. Energy Distribution of Slow Neutrons Scattered from Solids. Phys. Rev. 88:542 – 547, 1952.
23. Lowde R.D. Diffuse Reflection of Neutrons from a Single Crystal. Proc. Phy. Soc. A65:857 – 858, 1952. On the diffuse reflexion of neutrons by a single crystal. Proc. Roy. Soc. A221:206 – 223, 1954.
24. Brockhouse B.N. A New Tool for the Study of Energy Levels in Condensed Systems (A). Study of Energy Levels in Solids, Liquids and Magnetic Materials by Scattered Slow Neutrons (T). Phys. Rev. 98:1171, 1955.
25. Brockhouse B.N. Energy Distribution of Neutrons Scattered by Paramagnetic Substances. Phys. Rev. 99:601-603(L), 1955.26. Jacrot B. Mesure de l'energie de neutrons tres lents apres une diffusion inelastique par des polycrystaux et des monocrystaux. Compt. Rend. 240:745 – 747, 1955.
27. Brockhouse B.N. Neutron Scattering and the Frequency Distribution of the Normal Modes of Vanadium Metal. Can. J. Phys. 33:889 – 891, 1955.

28. Brockhouse B.N. and Stewart A.T. Scattering of Neutrons by Phonons in an Aluminum Single Crystal. Phys. Rev. 100:756 – 757, 1955. Normal Modes of Aluminum by Neutron Spectrometry. Rev. Mod. Phys. 30:236 – 249, 1958.
29. Carter R.S., Palevsky H. and Hughes D.J. Inelastic Scattering of Slow Neutrons by Lattice Vibrations in Aluminum. Phys. Rev. 106:1168 – 1174, 1957.
30. Larsson K.-E., Dahlborg U. and Holmryd S. A study of some temperature effects on the phonons in aluminium by use of cold neutrons. Ark. Fys. 17:369 – 392, 1960.
31. Eisenhauer C.M., Pelah I., Hughes D.J. and Palevsky H. Measurement of Lattice Vibrations in Vanadium by Neutron Scattering. Phys. Rev. 109:1046 – 1051, 1958.
32. Stewart A.T. and Brockhouse B.N. Vibration Spectra of Vanadium and a Mn-Co Alloy by Neutron Spectrometry. Rev. Mod. Phys. 30:250 – 255, 1958.
33. Brockhouse B.N. Structural Dynamics of Water by Neutron Spectrometry. Acta Cryst. 10:827 – 828, 1957. Suppl. Nuovo Cimento 9:45 – 71, 1958. Diffusive Motions in Liquids and Neutron Scattering. Phys. Rev. Letters 2:287-289, 1959. Sakamoto M., Brockhouse B.N., Johnson R.G. and Pope N.K. Neutron Inelastic Scattering Study of Water. Source E, Suppl. B – II:370 – 373.
34. Brockhouse B.N. and Pope N.K. Time-dependent Pair Correlations in Liquid Lead. Phys. Rev. Letters 3:259 – 262, 1959.Brockhouse (References) May 9/95) # 5 of 7 for "Les Prix Nobel" (1994)
35. Cohen M. and Feynman R.P. Theory of Inelastic Scattering of Cold Neutrons from Liquid Helium. Phys. Rev. 107:13-24, 1957.
36. Palevsky H., Otnes K. and Larsson K.E. Excitation of Rotons in He II by Cold Neutrons. Phys. Rev. 108:1346 – 1347, 1958. Phys. Rev. 112:11 – 18, 1959.
37. Yarnell J.L., Arnold G.P., Bendt P.J. and Kerr E.C. Energy vs Momentum Relations for Excitations in Liquid Helium. Phys. Rev. Lett. 1:9 – 11, 1958. Excitations in Liquid Helium: Neutron Scattering Measurements. Phys. Rev. 113:1379 – 1386, 1959.
38. Henshaw D.G. Energy-Momentum Relation in Liquid Helium in Inelastic Scattering of Slow Neutrons. Phys. Rev. Lett. 1:127 – 129, 1958. Henshaw D.G. and Woods A.D.B. Modes of Atomic Motions in Liquid Helium by Inelastic Scattering of Neutrons. Phys. Rev. 121:1266 – 1274, 1961.
39. Brockhouse B.N. Scattering of Neutrons by Spin Waves in Magnetite. Phys. Rev. 106:859 – 864, 1957. Field Dependence of Neutron Scattering by Spin Waves. Phys. Rev. 111:1273 – 1274, 1958.
40. Watanabe H. and Brockhouse B.N. Observation of Optical and Acoustical Magnons in Magnetite. Phys. Letters 1:189 – 190, 1962. Exchange field splitting in Ytterbium Iron Garnet. Phys. Rev. 128:67, 1962.
41. Sinclair R.N. and Brockhouse B.N. Dispersion Relation for Spin Waves in a fcc Cobalt Alloy. Phys. Rev. 120: 1638 – 1640, 1960.
42. Brockhouse B.N., Becka L.N., Rao K.R., Sinclair R.N. and Woods A.D.B. Crystal Field Spectra in Rare Earth Oxides, Source E, Suppl. B-III:63 – 66, 1962.
43. Brockhouse B.N. Methods for Neutron Spectrometry. Source C:113 – 150, 1961.
44. Svensson E.C., Brockhouse B.N. and Rowe J.M. "In-Band" modes of vibration of a dilute disordered alloy - Cu(Au). Solid State Comm. 3:245 – 249, 1965.

-PHONONS-

45. Brockhouse B.N. and Iyengar P.K. Normal Vibrations of Germanium by Neutron Spectrometry. Phys. Rev. 108:894 – 895, 1957. Normal Modes of Germanium by Neutron Spectrometry. Phys. Rev. 111:747 – 754, 1958.
46. Brockhouse B.N. Lattice Vibrations in Semiconductors by Neutron Spectrometry. J. Phys. Chem. Solids 8:400 – 405, 1959. Lattice Vibrations in Silicon and Germanium. Phys. Rev. Letters 2:256 – 258, 1959.
47. Brockhouse B.N. and Dasannacharya B.A. Temperature Effects on Lattice Vibrations in Germanium. Solid State Comm. 1:205 – 209, 1963.
48. Yarnell J.L., Warren J.L. and Wenzel R.G. Lattice Vibrations in Diamond. Phys. Rev. Letters 13:13 – 15, 1964.
49. Woods A.D.B., Cochran W. and Brockhouse B.N. Lattice Dynamics of Alkali Halide Crystals. Phys. Rev. 119:980 – 999, 1957.

50. Woods A.D.B., Brockhouse B.N. and Cowley R.A. and Cochran W. Lattice Dynamics of Alkali Halide Crystals. II. Experimental Studies of KBr and NaI. Phys. Rev. 131:1025 – 1029, 1963. Cowley R.A., Cochran W. and Brockhouse B.N. and Woods A.D.B. III. Theoretical. Phys. Rev. 131:1030 – 1039, 1963.
51. Brockhouse B.N., Arase T., Caglioti G., Rao K.R. and Woods A.D.B. Crystal Dynamics of Lead. I. Dispersion Curves at 100 K. Phys. Rev. 128:1099 – 1111, 1962.
52. Kohn W. Image of the Fermi Surface in the Vibration Spectrum of a Metal. Phys. Rev. Letters 2:393 – 394, 1959.
53. Brockhouse B.N., Rao K.R. and Woods A.D.B. Image of the Fermi Surface in the Lattice Vibrations of Lead. Phys. Rev. Letters 7:93 – 95, 1961.
54. Woods A.D.B., Brockhouse B.N., March R.H. and Stewart A.T. and Bowers R. Crystal Dynamics of Sodium at 90 K. Phys. Rev. 128:1112 – 1120, 1962.
55. Dixon A.E., Woods A.D.B. and Brockhouse B.N. Frequency Distribution of the Lattice Vibrations in Sodium. Proc. Phys. Soc. 81:973 – 974, 1963.
56. Gilat G. and Dolling G. Normal Vibrations of β Brass. Phys. Rev. 138:A1053 – 1065, 1965.
57. Stedman R. and Nilsson S. Phonons in Aluminium at 80 K. Source D:Vol.I:211 – 223, 1965.

-REVIEWS-

58. Shull C.G. and Wollan E.O. Applications of Neutron Diffraction to Solid State Problems. Source H:Vol.2:138 – 217, 1956.
59. Kothari L.S. and Singwi K.S. Interaction of Thermal Neutrons with Solids. Source H:Vol.8:110 – 190, 1959.
60. Brockhouse B.N. Crystal and Liquid Dynamics from Neutron Energy Distributions. Source G:129 – 144, 1962. Phonons and Neutron Scattering. Source I:221 – 275, 1964. Neutron Scattering by Phonons. Source J:110 – 152, 1966.
61. Brockhouse B.N., Hautecler S. and Stiller H. Inelastic Scattering of Slow Neutrons. Source F:580 – 642, 1964.
62. Dolling G. and Woods A.D.B. Thermal Vibrations of Crystal Lattices. In Source K, editor P.A. Egelstaff. K:193 – 249, 1965.
63. Lomer W.M. and Low G.G. Introductory Theory. In Source K, editor P.A. Egelstaff. K:1 – 52, 1965. Other articles by R.M. Brugger, P.K. Iyengar, S.J. Cocking and F.J. Webb, G. Dolling and A.D.B. Woods, B. Jacrot and T. Riste, A. Sjolander, K.E. Larsson, J.A. Janik and A. Kowalska provide rather complete reports on the state of the field by 1965.

Clifford G. Shull

CLIFFORD G. SHULL

I was born on September 23, 1915 to my parents, David H. and Daisy B. Shull, in the section of Pittsburgh, Pennsylvania, known as Glenwood, which obviously relates to their selection of my middle name. I was preceded by an older sister, Evalyn May, and an older brother, Perry Leo, so that I grew up as the baby in the family. Both my father and mother had origins in rural, central Pennsylvania, in farming sections of Perry County. After moving with his then family to the big city, Pittsburgh, my father started a small business that evolved into a hardware store and an associated home repair service.

My early years of growth were entirely normal and happy ones and I had the usual collection of friends and buddies, who were often seen on the ball field or on roller skates. Grade schooling was nearby, a few blocks from our home, and this led later on to junior high school in the adjoining Hazelwood section but still within walking distance of our home. Following this, I had decided to go to Schenley High School for the remaining three years of school work and this required a more troublesome commute of 45 minutes by public street car. My first interest in physics as a career speciality came during my senior year at Schenley when I took the physics course taught by Paul Dysart. Somewhat older than the usual high school teacher and with a PhD degree in his background, he was a very impressive teacher who delighted in demonstrations from his laboratory and in explaining the principles behind them. Thereafter my original interest in aeronautical engineering was in heavy competition with physical science.

It seemed natural, in view of limited family financial straights, that I should continue into college study by living at home and commuting to the Carnegie Institute of Technology (now Carnegie Mellon University). Carnegie Tech was also located in the Schenley Park district of Pittsburgh so that essentially the same commute was called for and it offered good, reputable curricula in the engineering and physical sciences. I was pleased when offered admission to the fall term of 1933 and particularly so when given a half-tuition scholarship in view of my good high school record. Once there, my interest in physics as a major subject sharpened quickly, helped along no doubt by the brilliant lectures in my freshman physics course given by Harry Hower, the chairman of the Physics Department. Hower was more aptly labeled an optical and illuminating engineer than a physicist, because of his extensive consulting activities in coastal lighthouse lens design and other architectural problems, but his lectures were delightful, inspiring and not often-to-be-missed by his students.

Shortly after my admission at Carnegie Tech, a family crisis developed when my father died unexpectedly in January, 1934. By this time, my sister had married and, with her husband, were living at home along with my brother (who had just finished college as an art major), my mother and myseelf. My brother decided to forego his art teaching and operate my father's business and this continued until I had finished Carnegie Tech in 1937. The four years spent there were entirely pleasurable ones, in spite of the time-consuming commute, and I enjoyed the association with my fellow students and professors in the department. I was able to work in the summer periods at jobs both on and off campus and this helped to meet my rather minimal expenses during the year. Among the professors, I valued very much the friendly encouragement and counsel offered by Emerson Pugh during my junior and senior years, leading to my continuance into graduate school at New York University in the fall of 1937.

New York University was then a very large university, perhaps the largest in the nation, with several distributed, more or less autonomous, campuses. I was located with the Physics Department at the University Heights campus in the upper Bronx section of New York City and my teaching assistantship provided living subsistence, teaching meaning laboratory course help and problem assignment grading. We graduate students were encouraged at an early stage to join and help in one of the half dozen or so ongoing research projects within the department. I became associated with a nuclear physics group headed by Frank Myers and Robert Huntoon, who were in the process of building a 200 keV Cockroft-Walton generator for accelerating deuterons. Much valuable experience was obtained with this exposure by Craig Crenshaw, another graduate student, and myself and we were able to help in the initial experiment with this accelerator, a study of the D-D nuclear reaction.

During the third year of my graduate study, the Department decided that it could support the construction of a new 400 keV Van de Graaff generator to be used for accelerating electrons. Frank Myers took on this responsibility with me as his assistant and the thought that it could be used to repeat the electron-double-scattering (EDS) experiment as a possible thesis topic for me. This EDS type of experiment loomed important at the time because it was considered a direct test that electrons have a spin or polarization. Several earlier experiments had given either negative or inconclusive results and it seemed worthwhile that the experiment be performed again under new conditions. The construction and testing of the new facility went smoothly and I turned to getting ready for my thesis EDS experiment. By this time, Frank Myers had decided to take his overdue sabbatical leave with Robert Van de Graaff at MIT. I was fortunate in getting Richard Cox, a senior professor in the department, to supervise and offer expert and friendly advice on my efforts. Finally after four months of data collection and analysis, the experiment was successful and I was able to prepare a thesis and take my PhD degree in June 1941.

Among the other research programs being pursued by the NYU department was the study of neutron interactions with materials as started by Alan Mitchell and carried on by Martin Whitaker. Using a Ra-Be neutron source surrounded by a paraffin howitzer, a modest beam of thermalized neutrons was available for experimentation and, during my period at the Heights, this was directed towards a search for the expected paramagnetic scattering from certain materials. Theoretical prediction of this had been given by O. Halpern and M. Johnson and their students in the Department. I was familiar with this problem through my contemporary graduate student William Bright who worked with Whitaker on the experiment and indeed found myself working on the same problem a decade later.

I have neglected to mention an important event that occurred in my first year in New York City. Through my good friend Craig Crenshaw, I was introduced to a young lady, Martha-Nuel Summer, who had recently come from South Carolina to the graduate school at Columbia University to study early American History. Our friendship flourished during the years of our professional studies and we married shortly after I took my degree and had a job in waiting. She has remained my loving companion to the present and along the way we have been favored by three fine sons, John, Robert and William, who have beautiful families of their own.

I had arranged for a position at Beacon, NY with the research laboratory of The Texas Company, and Martha and I set up housekeeping there in July 1941. This laboratory addressed problems associated with the production and use of petroleum fuels and lubricants and included a small group of physicists. I was asked to study the microstructure of catalysts using gas adsorption and x-ray diffraction and scattering as tools for characterizing the physical structure of these materials. These catalysts were used in the production of high-performance aviation fuel and this area of investigation became increasingly important after the US entry in the World War in December 1941. Of singular significance to the scientific community in the first year of our wartime activity was the growth of the Manhattan Project dealing with the development of an atomic weapon. Many scientists had been drawn into this, including a number of my old colleagues and professors from graduate school. I was encouraged to join them and would have done so except that The Texas Company would not agree to my wartime job change. The matter was finally settled in their favor by an adjudication hearing at an area manpower board and I stayed in Beacon through the war years.

My work at Beacon was interesting and challenging and gave me the opportunity of learning things about diffraction processes, crystallography and the new field of solid state physics. Through visits and early meetings of the American Society for X-ray and Electron Diffraction, I was able to know established personages such as Warren, Buerger, Fankuchen, Zachariasen, Ewald, Harker, Gingrich and Donnay. Once the war was over, my interest in participating in the exciting new developments in nuclear physics within the Manhattan Project returned, and I paid a visit to the Clinton Laboratory (now Oak Ridge National Laboratory) in Tennessee. The activity there fascinated me very much and I convinced Martha that we should move there, which we did in June 1946 along with our one and a half year old son.

It was arranged that I would work with Ernest Wollan, who had been at the Laboratory since its formation during the war period and who had just assembled a rudimentary two-axis spectrometer for obtaining neutron diffraction patterns of crystals and materials. Wollan had shown me his first powder diffraction pattern on my earlier visit and I was delighted to be able to join him in exploring how neutron patterns could be used to supplement those obtained with x-rays or electrons. Our collaboration on common problems was to continue for nearly a decade until I left Oak Ridge in 1955 for academic life at Massachusetts Institute of Technology. I regret very much that Wollan's death in 1984 precluded his sharing in the Nobel honor that has been given to Brockhouse and me since his contributions were certainly deserving of recognition.

I was attracted to MIT by the prospects of teaching and of training graduate research students at the soon-to-be-completed MITR-I research reactor on campus. This reactor was among the early group of condensed volume reactors using isotopically enriched fuel which were being explored in that period. Together with occasional post-doctoral students and a regular flow of graduate thesis students, our group carried on investigations using neutron radiation from this reactor in many fields until my retirement from MIT in 1986. These studies included: internal magnetization in crystals, development of polarized beam technology, dynamical scattering in perfect crystals, interferometry, and fundamental properties of the neutron. The opportunity of being at MIT with its fine faculty and excellent students has certainly been most stimulating and satisfying.

EARLY DEVELOPMENT OF NEUTRON SCATTERING

Nobel Lecture, December 8, 1994

by

CLIFFORD G. SHULL

Department of Physics, Massachusetts Institute of Technology, Cambridge, MA 02139, USA

Neutrons were discovered by Chadwick in 1932 when he observed a penetrating form of radiation emanating from beryllium metal when activated by alpha-particles from a radium source. Further study showed this to be neutral particle radiation which could be degraded in kinetic energy to thermal energy upon successive inelastic scattering by light atoms in a medium. With this Fermi thermalization, it was realized that thermal neutrons, because of the wave-particle duality principle, should exhibit a wave character with DeBroglie wavelength comparable to the atom spacing in solids. Thus diffraction effects should be expected in the scattering of neutrons by crystals, just as with x-radiation, and early experiments in 1936 showed in a crude way that this was true.

At this time, x-ray diffraction had been developed to become an important tool in establishing the structure and atomic interactions in materials and it seemed unlikely that neutron diffraction would develop as a useful tool because of very low source intensity. This limitation changed dramatically in the early war years of 1939 – 1943 with the discovery of nuclear fission by Hahn and Meitner and the subsequent demonstration by Fermi of a self-sustaining, neutron chain-reacting assembly. Following this Fermi demonstration, immediate design and construction of a full scale nuclear reactor, or "pile" as it was called then, was effected at Oak Ridge, Tennessee.

This reactor, then called the Clinton Pile, was meant to produce the first measurable quantities of plutonium and to serve as a pilot plant for the much larger production reactors which were being designed for construction in the state of Washington. The Clinton Pile was a graphite moderated, air-cooled assembly which operated at a power level of about 3.5 megawatts, thereby producing a slow neutron flux density of about 10^{12} neutrons/cm^2 sec. During this same period, another assembly was designed and constructed at Chicago with the feature of heavy water moderation, the CP-3 assembly. I show in Fig. (1) the start-up dates of these early and post-war reactors. Both the Clinton and the CP-3 pile assemblies operated through the war years producing man-made elements and isotopes and to a limited extent their neutron radiation was used to obtain some critically needed cross section data.

Fig. (2) shows an early photograph of the loading face of the Clinton Pile with operators performing some function on a movable platform in front of the assembly. The pile is cubic in form with edge length about 7 m and consists of a nestled collection of graphite "stringers", rectangular blocks of square cross sections about 10 cm on a side and length 150 cm. Pure uranium metal serves as the fissionable fuel for the assembly and is positioned in a lattice-array of channels in the graphite extending from the loading face to the back, discharge face. The uranium metal is in the form of cylindrical rods, about 2 cm in diameter and 10 cm in length, and each is carefully sealed in an aluminum can.

Start–up Dates of Early "Piles"

1942 (*Dec.*)	Fermi Demonstration (*Chicago*)		
1943 (*Nov.*)	Clinton Pile, graphite, (*3 MW*)		
1944	Argonne CP–3, heavy water		
1947	NRX	–	Chalk River
1948	BEPO	–	Harwell
1950	LITR	–	Oak Ridge (enriched U)
	BGRR	–	Brookhaven

Fig. 1: Dates of Operation of Early Nuclear Reactors.

Fig. 2: Photograph of the Loading Face of the Clinton Pile.

Cooling air is drawn through the channels to the back face by large suction fans and control of the nuclear reaction level is effected by the positioning of various neutron-absorbing control rods within the assembly. Surrounding the cubic graphite pile on all sides is a radiation protection shield of concrete of thickness about 1.5 m. With foresight, the designers had arranged some channels through the concrete shield for access to the active volume of the assembly.

During the war period, a single-axis spectrometer had been used at one of these channels to supply monoenergetic neutrons for transmission cross section measurements. At the end of the war Ernest Wollan decided to set up a two-axis spectrometer for full diffraction pattern study with the first axis providing a monochromatic beam, hopefully of enough intensity to permit second axis diffraction patterns to be studied. Wollan had arranged for delivery to Oak Ridge of a base spectrometer that he had used in his earlier thesis work on x-ray gas scattering under Arthur Compton at Chicago. This spec-

Fig. 3: Photographs of the First Double Axis Neutron Spectrometer Used to Take Powder Diffraction Patterns at the Clinton Pile.

trometer, with coaxial control of specimen and detector positions, had to be modified to support the larger loading of a neutron detector with its necessary shielding, and this was done with support cables running to a swivel-bearing in the ceiling directly above the spectrometer axis. Fig. (3) is a photograph of this assembly taken during the first year of use but after some improvements in the shielding around the detector and the monochromating crystal had been made.

Operation of the spectrometer was entirely by hand and it was a time-consuming chore for Wollan and his early colleague, R. B. Sawyer, to collect what were the first neutron diffraction patterns of polycrystalline NaCl and light and heavy water in the early months of 1946. I was shown these patterns on a visit that I made to Oak Ridge in the spring of that year prior to my relocating there and joining Wollan in further work in June 1946. I show in Fig. (4) a portion of the NaCl pattern and, although the background level is high, it is seen that the diffraction peaks can be measured and interpreted in a quantitative way. This was important because Wollan had hoped by using powder-diffraction patterns to avoid the uncertainties in single-crystal intensities caused by extinction effects. The light and heavy water patterns were taken to see if there was any suggestion of hydrogen (deuterium) atom contribution to the diffuse, liquid patterns as taken with neutrons. Some differences in the coherent features of the two patterns were to be seen but no attempt to analyze this was made.

During the early period of study with this spectrometer the emphasis was centered on understanding the intensity that was being measured in the diffraction pattern, both within the coherent diffraction peaks and in the diffuse scattering. At that time, the only pertinent, quantitative information that was available were values of the total scattering and absorption cross sections as measured with thermal neutrons for many elements. Although these quantities might be expected to establish the total intensity in all of the diffraction pattern, the existence of isotopes for some elements, the possible presence of different spin state scattering amplitudes for nuclei with spin, and the presumed thermal disorder or excitation scattering were all factors that had to be considered case by case. Progress in isolating and defining these various factors came very slowly because of the choresome mode of data collection on the instrument, which was entirely by hand operation.

Towards the end of the first year, we were joined by a young technician, Milton Marney, who assisted in the tedious data collection and by two volunteer physicist-scientists from the Oak Ridge Training School, William Davidson and George Morton. The Training School was organized in late 1946 and was meant to bring a blue-ribbon group of established scientists from academia, industry and government to Oak Ridge for an extended period so as to attend lectures, collaborate with research groups and generally to learn the new technology and science that had been developed during the war period. Davidson and Morton took interest in our neutron scattering work and helped with the data collection: however, Morton, who was skilled in instrumentation, soon tired of this and decided to design an automatic control system that would permit step-scanning with intensity measurement and its recording. This took the form of motor drives, cam wheels and microswitches, and most importantly a print-out recorder of the neutron counter measurement. These changes took several months to effect and, once operational, it permitted continuous, uninterrupted data collection and this represented a very important step in the progress of the program.

Along with these improvements in the instrumentation, regular changes were being made in the shielding around the monochromating crystal and the movable detector on the outer spectrometer. Initially, the background intensity in the diffraction patterns was comparable to the peak intensity, as seen in Fig. (4), and about as much time was spent in measuring background as in studying the pattern itself. Regular background intensity measurements through insertion of a solenoid-activated cadmium filter into the incident beam were a part of the initial automatic program system.

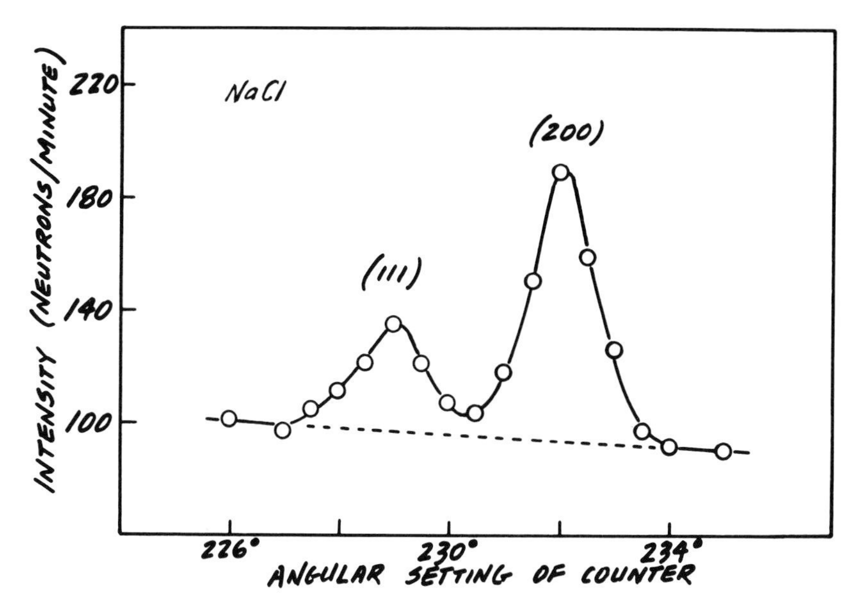

Fig. 4: Portion of the First Powder Difraction of NaCl taken at the Clinton Pile.1

With this new system of data accumulation, we returned to our initial program of classifying various elements according to their coherent scattering amplitudes and of exploring various novel cases where neutron diffraction might supply useful information beyond that obtained with x-ray scattering. In the special case of a monoisotopic element with zero nuclear spin (nearly satisfied in oxygen and carbon), the coherent scattering amplitude, which defines the coherent peak intensity in the diffraction pattern, can be obtained from the total scattering cross section. This can then be used as an internal standard in quantifying the diffractive properties of other elements when combined with the standard element. This program of classification went on for several years continuing into the period when separated isotopes became available.

Rather early in the program, a puzzling feature of the intensity distribution was noted, namely that there was larger than expected diffuse scattering in the pattern. Careful intensity measurements on an absolute scale showed the diffuse intensity to be larger than that caused by thermal diffuse scattering or isotopic and nuclear spin incoherence. This was eventually identified as resulting from multiple scattering within the scattering samples, a novel effect characteristic of neutron scattering and never encountered in equivalent x-ray scattering technology.

With these improvements in instrumentation and a better understanding of the components that were being seen in the patterns, attention was then turned to the important crystallographic case of hydrogen-containing crystals. It was known at the time that the coherent scattering amplitude of hydrogen for neutrons must be small since that quantity could be derived from measurements of total scattering of neutrons by ortho- and parahydrogen. However the precision of those measurements did not specify either its finiteness or its algebraic sign. There was also the additional suspicion that hydrogen atoms might display anomalously large thermal motion in crystals and that this would wipe out coherent scattering. This all became settled when powder patterns of NaH and NaD were obtained, with parts being shown in Fig. (5). Analysis of the intensities in these full patterns yielded the coherent scattering amplitudes for hydrogen and deuterium (they are of opposite sign) and showed the hydrogen centers to possess rather normal

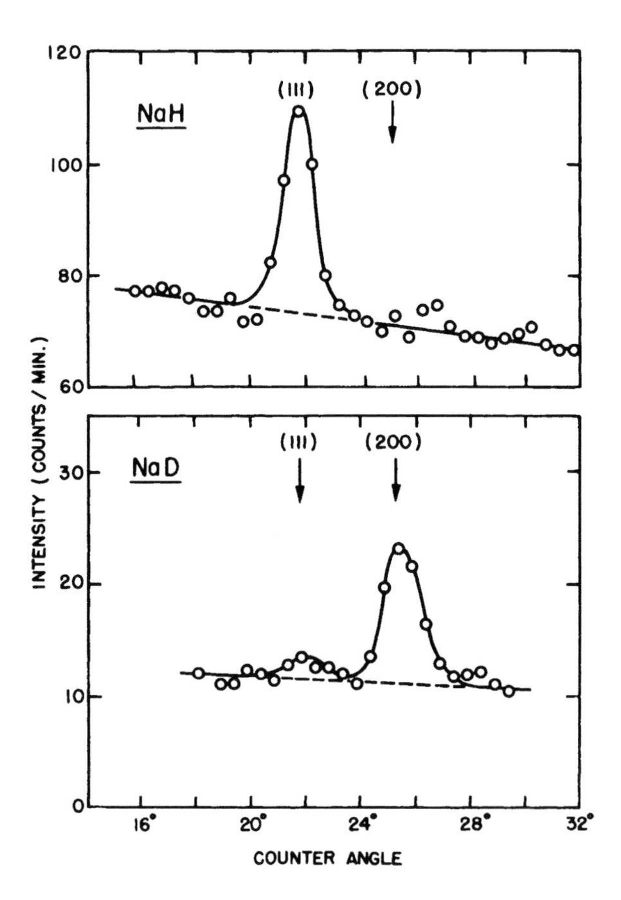

Fig. 5: Patterns for NaH an NaD Showing that Hydrogen Contributes Normally to the Pattern.

thermal oscillation. The large diffuse scattering seen for NaH arises from anomalously large nuclear spin incoherent scattering from hydrogen centers.

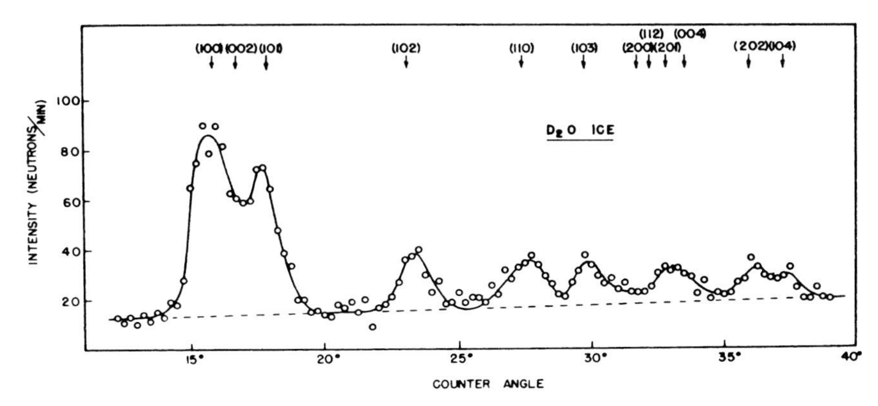

Fig. 6: Pattern Taken for a Sample of Powdered D_2O Ice.

Another interesting hydrogen structure that attracted our attention was that of common ice. It was known from x-ray diffraction studies that the basic structure was formed by interlocking oxygen atom tetrahedra but the bonding location of the hydrogen atoms was unknown and several models for this had been proposed. Fig. (6) shows a powder pattern that was taken with a deuterated sample of ice and the analysis of it shows clearly the validity of the

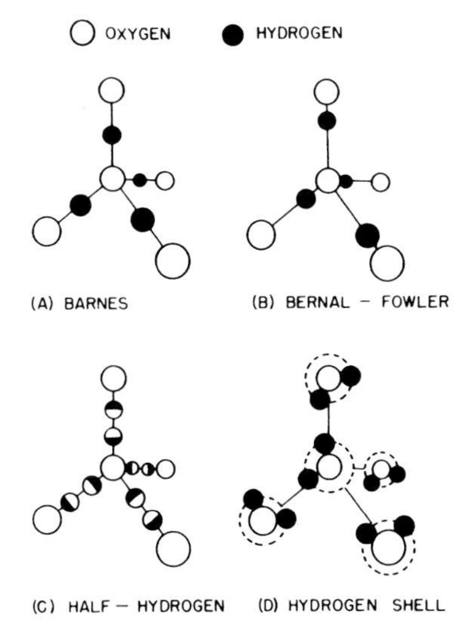

Fig. 7: Suggested Models for the Hydrogen Atom Distribution in Ice.

"half-hydrogen or Pauling" model of the structure as illustrated in Fig. (7) Preparing a powder sample of D_2O ice presented new problems and a new specimen cryostat (with dry ice-acetone coolant) was built which some people called an "ice-ball machine". D_2O vapor was frozen on the inside surface of a cylindrical specimen tube, followed by reaming and light packing of successive layers without contamination from normal atmospheric moisture. Other hydrogen structures were studied in the next two years, mostly transition metal and heavy atom hydrides including the interesting, extreme case of uranium hydride.

In mid-1949 a second spectrometer was installed at a beam port adjoining the early one. This spectrometer, again a two-axis unit, was really the first prototype of present-day neutron spectrometers, having components that were built for specific neutron use rather than being improvised from x-ray scattering units. An elegant cylindrical shield (A) surrounded the monochromating crystal and this was supported on a large bearing which permitted easy change of beam exiting angle or beam wavelength. The specimen spectrometer (B) was supported from the monochromator shield and was now sturdy enough to support the counter arm without the use of support cables that characterized the earlier unit.

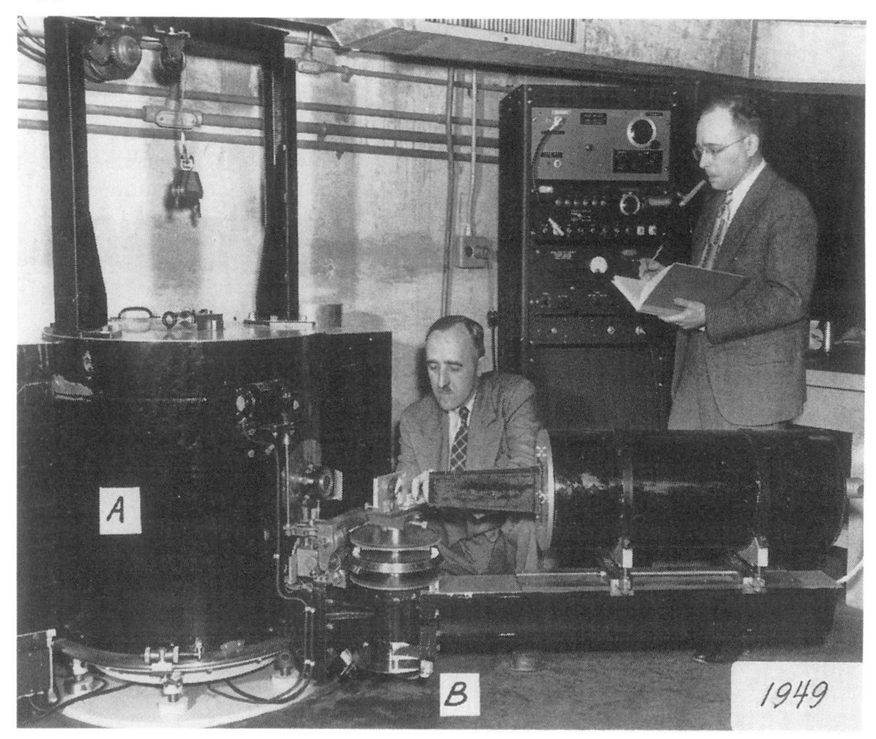

Fig. 8: Shows this spectrometer with Ernest Wollan (kneeling) and the author in an early photograph.

Along with the use of these spectrometers in which neutron-sensitive counting tubes were used to detect diffraction intensity effects, we were interested in exploring photographic detection as used effectively in x-ray technology. It was realized that photographic film would be very insensitive to neutron radiation and that a converter screen placed next to the emulsion would be required. After some tests, it was found that an indium sheet would serve this purpose: indium has sizable absorption cross sections for thermal neutrons and emits beta-particles in its decay which will produce photographic action. This screen-film detector was used to record the first neutron Laue pattern, taken for a NaCl crystal with an exposure time of about 10 hours as shown in Fig. (9). A direct, full spectrum beam was taken from the reactor through a long shielding collimator and this passed on to the crystal with the film positioned about 10 cm behind the crystal. Among the features to be seen in this first pattern is the apparent double structure of the Laue spots. This puzzled us at first, until later investigation showed it to be an artifact caused by twinning in our first selected crystal. One can notice also in the background, a series of narrow band images running from top to bottom. This results from the presence of Scotch Tape ribbon holding adjacent strips of indium sheet together. So not only is this the first neutron Laue pattern but, as well, it is the first neutron radiograph!

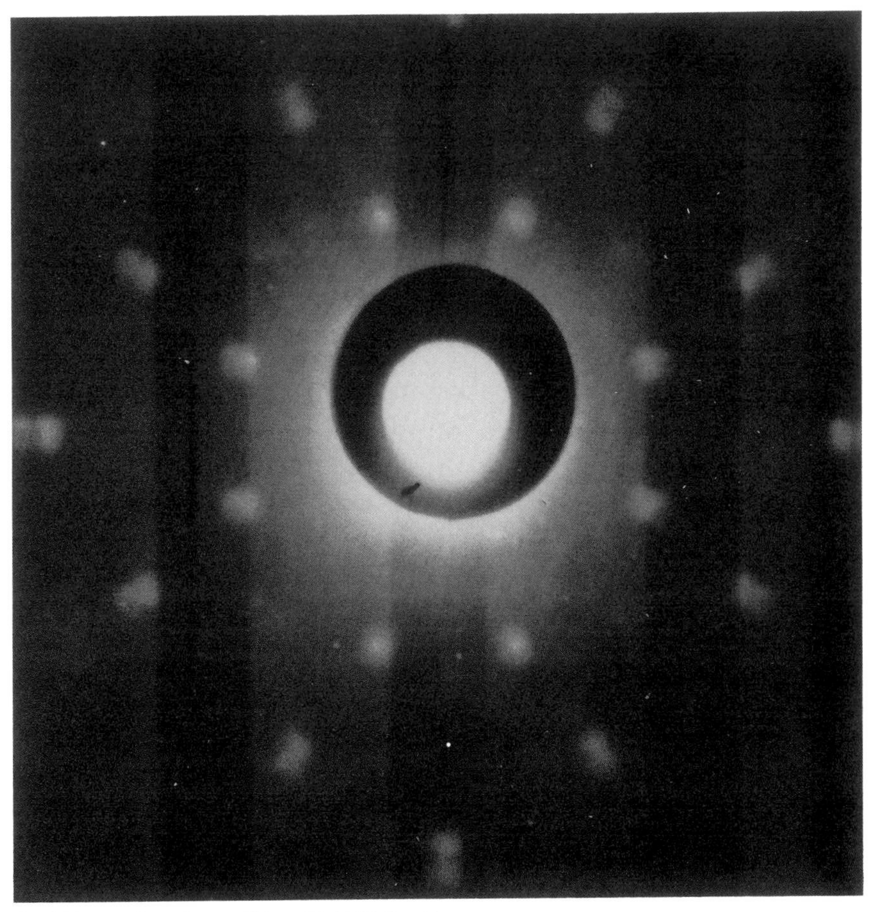

Fig. 9: Neutron Laue Pattern Taken with NaCl Crystal.

My abbreviated description of the developments that occurred during the first few years of neutron scattering investigation at Oak Ridge has failed to mention that many other colleagues and collaborators beyond those specifically mentioned were involved in the effort. My heartfelt appreciation is due them as well as to other members of the Physics Division and the Oak Ridge laboratory administration for unfailing support. But mostly I appreciate the association, collaboration and close friendship, extending for nearly a decade, that I enjoyed with Ernest Wollan, who first guided me to the wonders of neutron scattering research. I regret that he did not live long enough to share in the honors that have come to me.

Physics 1995

MARTIN L. PERL

for the discovery of the tau lepton

and

FREDERICK REINES

for the detection of the neutrino

THE NOBEL PRIZE IN PHYSICS

Speech by Professor Carl Nordling of the Royal Swedish Academy of Sciences.
Translation of the Swedish text.

Your Majesties, Your Royal Highnesses, Ladies and Gentlemen,

Physicists believe that all matter, for example the matter in our own bodies, consists of quarks and leptons. Quarks are heavy, and leptons are light. There are two types of quarks, which serve as the building blocks of atomic nuclei. Leptons, which occur outside of atomic nuclei, also come in two types: electrons, which have both an electrical charge and a measurable mass, and neutrinos, which lack both charge and mass. This one quark-lepton family, with its four members, is enough to account for all the matter in the universe today.

But the universe has a long history behind it. In its early stages, conditions were entirely different from today. There were very high temperatures and very high concentrations of energy. In such a climate, other quark-lepton families also thrived. With the aid of accelerators, physicists have re-created for brief moments the extreme conditions under which a second and even a third quark-lepton family can appear. But that was all. Physicists have shown that there is no fourth family of quarks and leptons within the existing paradigm of elementary-particle physics.

A long series of discoveries underpins the three-family model. Two of these discoveries are being honored with this year's Nobel Prize in Physics. Both concern the discovery of leptons: one belonging to the first quark-lepton family, the other to the third such family. Both discoveries provide answers to deep, fundamental questions in physics.

Together with the late C.L. Cowan, Frederick Reines detected the neutrino, the sister lepton of the electron in the first quark-lepton family, even before the family concept had emerged. It was a long-awaited discovery. For nearly 25 years, physicists had been waiting for someone to accomplish this feat. Meanwhile the neutrino had been a mental construct that physicists had needed in order to "save" the law of conservation of energy in certain types of radioactive decay. But it seemed impossible to verify the neutrino's actual existence. It flashed undetected past every observer, at the speed of light.

Reines realized that a nuclear reactor must emit copious quantities of neutrinos, although no one had noticed them before. During the 1950s, he and Cowan developed a method for capturing at least a few of these elusive subatomic particles.

After an unsuccessful trial, they devised a modified experiment that yielded favorable indications. A couple of years later, their findings were unequivocal. They had proved the existence of the neutrino. This discovery was a milestone of modern physics. It opened the way to a major new field of research, neutrino physics. Reines has also played a vital role in subsequent phases of this work.

There were now three known leptons. First was the electron, which had been discovered as early as 1897. Second was the newly discovered neutrino. Third was the muon, a heavier version of the electron and the odd man out among elementary particles. No quarks had entered the picture yet.

The idea that there might be a third, extremely heavy relative of the electron emerged during the 1960s, but such a particle fell outside the theoretical framework accepted at that time. The difficulties of experimentally proving the existence of such a heavy lepton also seemed almost insurmountable. But for many physicists, such almost insurmountable barriers exert an attraction all their own.

At Stanford, Martin Perl planned an experiment that might possibly resolve the matter. He needed a sufficiently strong energy source, and Stanford had the world's most powerful accelerator for this purpose. He also needed to figure out in what way the new lepton might reveal its presence. There were few leads, however, and great inventiveness was required to design an experiment that would accomplish this.

When Perl and his coworkers wrote in 1975 that they had found a third lepton, nearly four thousand times heavier than the electron first identified in 1897, they were announcing a major discovery. They had identified the first member of the third and final quark-lepton family.

The tau, or tau lepton, as this super-heavy cousin of the electron was designated, provided the key to a definitive understanding of the family structure of elementary particles. Some day the tau neutrino, its sister lepton, may prove to account for a large portion of the missing mass in the universe and to play an important role in supernova explosions, and thus in cosmology. The tau lepton itself may be of crucial importance in testing future theories of how matter acquired the property that we call mass.

Professor Perl, Professor Reines,

You have been awarded the 1995 Nobel Prize in Physics for your outstanding contributions to lepton physics. It is a privilege and a great pleasure for me to congratulate you on behalf of the Royal Swedish Academy of Sciences, and I now ask you to receive your Nobel Prizes from the hands of His Majesty the King.

Martin L. Perl

MARTIN L. PERL

Good Schools, Books, a Love of Mechanics, and You Must Earn a Living

About 1900 my parents came to the United States as children from what was then the Polish area of Russia. As Jews, their families left Russia to escape the poverty and the antisemitism. My parents grew up in poor areas of New York City, my father Oscar Perl in the East Side district of Manhattan and my mother Fay Rosenthal in the Brownsville district of Brooklyn. Their educations ended with high school–my father going to work as a clerk and then salesman in a company dealing in printing and stationary, and my mother working as a secretary and then bookkeeper in a firm of wool merchants.

My parents were determined to move into the middle class. By the time my sister, Lila Perl, and I were born in the 1920's, my father had established a printing and advertising company called Allied Printing. For many years, Allied Printing was a precarious business. I remember conversations at the dinner table about the problem of meeting the upcoming Friday payroll. However, Allied Printing brought the four of us into the middle class and kept us in the middle class thru the Depression of the 1930's. We lived in the better neighborhoods of the borough of Brooklyn, not the fanciest neighborhoods, but quite good neighborhoods, and so we went to quite good schools.

These schools and the attitude of my parents towards these schools were important in preparing me for the work of an experimental scientist. Going to school and working for good marks, indeed working for very good marks, was a serious business. My parents regarded school teachers as higher beings, as did many immigrants. School principals were gods to be worshiped but never seen by children or parents. Parents never visited the school to talk about the curriculum or to meet with their child's teacher. A parent being called to the school because their child had misbehaved was as serious as a parent being called to the police station because their child had robbed a bank. The remoteness of my parents from the schools, so unfashionable today, was often painful for me, but I learned early to deal with an outside and sometimes hard world. Good training for research work! The experimenter dealing with nature faces an outside and often hard world. Natures' curriculum cannot be changed.

The curricula were unsophisticated, with a great deal of time wasted on penmanship and geography in the early grades and repetitions of the trivial history of New York City in higher grades. But there were also serious courses.

In my high school, two foreign languages had to be studied, four years of English was required, and that meant mostly grammar and composition. I was able to take four years of mathematics and a year of physics. Whatever the course, whether the course was boring or interesting to me, whether I was talented in mathematics or not talented in languages, my parents expected A's. This was good training for research, because large parts of experimental work are sometimes boring or involve the use of skills in which one is not particularly gifted.

For example, I am not a good draftsman. Until recently when I could use computer-based drafting programs, my drawings always looked messy, with uneven lines and ragged lettering. I could never get an "A" in drafting in college. Yet drawing the apparatus to be built for my experiments has always been a crucial part of my experimental work.

There was compensation for the unsophisticated curriculum; with good marks one could "skip" school years. The normal progression was to begin the eight years of elementary school at six years of age, and then to take four years of high school, leading to graduation at eighteen. But classrooms were crowded, and there were no worries about the proper social level of a student; a good student could skip a year or more in elementary school. I was sixteen when I graduated from James Madison High School in Brooklyn in 1942. My sister, who is now a well known writer in the United States, moved through school even faster–she graduated at fifteen and one-half.

Along with my parents insistence, soon internalized, that I do very well in school, went my love of reading and my love of mechanics. I read everything: fiction, history, science, mathematics, biography, travel. There were two free public libraries within walking distance of my home; I remember taking six books home from every visit, the limit set by the library. This reading had only partial approval from my parents. They wanted me to play more sports because they were acutely sensitive to their children being one hundred percent American, and they believed that all Americans played sports and loved sports. They felt that too much reading interfered with my going outside to play sport. I loved rainy days when I did not have to go outside, and to the present I still feel very content on a rainy day.

Two books are burned in my memory, Lancelot Hogben's Mathematics for the Millions and his Science for the Citizen. I borrowed them from the library again and again. I made summaries of them. I could not understand Hogben's introduction to calculus so I copied that section completely. I don't know why it never occurred to me or my parents to buy the books. We could have well-afforded them, but somehow buying books was a waste of money. Naturally, I have compensated in my adult years by owning very large numbers of books.

Another thing we could have afforded was to buy me an Erector construction set. The Erector construction set was the United States equivalent of Meccano or Märklin construction sets in England and Europe. But the cousin I played with every Saturday had an Erector set, and one Erector set per extended family was considered quite enough. He also had electric trains. I loved to build with the Erector set, I loved to build toys and models out of wood, I loved to draw mechanical devices, even those I could not build. I loved to read the magazines, *Popular Mechanics* and *Popular Science.* I loved all things mechanical; cars, trucks, derricks trains, and steam ships. I was in love with mechanics, and I still am.

Before leaving this subject I must mention that since I never owned an Erector set as a child, I have compensated in my adult years by collecting old European, English, and American construction sets; and even by devising and starting prototype production of a modern wooden construction set called BIG-NUT.

I was also interested in chemistry, but my parents were not willing to buy me a chemistry set. I had some chemicals but when I bought sulfuric acid and nitric acid, my father confiscated the acids on the grounds of safety. As every child knows, chemistry with nothing stronger than vinegar soon becomes dull. Strangely for a person who became a physicist, I was not interested in amateur radio or in building radios. I don't know why. This was the 1930's when vacuum tubes and variable condensers made radio building quite mechanical.

In spite of very good school marks, a love of books (particularly in science and mathematics), and a great love of mechanics, I never thought of becoming a scientist. That was because as the children of immigrants, my sister and I were taught that we must use our education to "earn a good living". In fact, we didn't have to be taught that. It was obvious to us. Our home life was physically comfortable, and in some ways emotionally supportive, but it was also rigid and stifling. We knew that we had to earn our own livings to escape from home and Brooklyn. A good living in the Jewish middle class meant that a girl should become a teacher or nurse; a boy should become a doctor, dentist, lawyer, or accountant. I did not think about going into business because the difficulties of the Depression years did not make business a good way to earn a living.

Although I won the physics medal when I graduated from high school, I did not think of becoming a physicist or any kind of scientist. My parents and I knew about a few scientists, certainly Pasteur, and perhaps Einstein, but we did not know that it was possible for a man to earn a living as a scientist.

Engineering Studies, the War, a Practicing Engineer, and What You are Interested in is called Physics

We did know that a man could earn a living as engineer. And so in choosing a profession for me, my parents and I took into account my love of mechanics, and science and mathematics. We put aside my becoming a doctor, dentist, lawyer, or accountant in favor of my becoming an engineer. This was an unusual choice for a Jewish boy in the early 1940's because there was still plenty of antisemitism in engineering companies. I enrolled in the

Polytechnic Institute of Brooklyn, now Polytechnic University, and began studying chemical engineering.

There were several reasons for choosing chemical engineering. Chemistry was a very exciting field in the late 1930's and early 1940's. Chemistry was bringing to our lives synthetic materials such as nylon. The slogan of the radio program, *Dupont's Cavalcade of America,* was "Better things for better living through chemistry". Furthermore, Allied Printing had prospered through my father's hard work, and through the inclusion of a few chemical companies among his customers. He became friends with buyers in several of these companies, and they told him about the expansion of their companies. There would always be a good job in chemical engineering.

One of the first courses I took in college was general physics, using the textbook by Hausman and Slack. The course was all about pulleys and thermometers; physics seemed a dead field compared to chemistry. So, just as I was blind to the fascination of physics in high school, I was once again blind to its fascination in college. I ignored physics, and continued studying chemistry and chemical engineering.

My studies were interrupted by the war. I wanted to join the Unıted States Army, but I was not yet eighteen and my parents would not give me permission. However, they agreed to me joining the United States Merchant Marine, I was allowed to leave college and become an engineering cadet in the program at the Kings Point Merchant Marine Academy. The training ship was wonderful–it had a main reciprocating steam engine, and direct steam driven pumps and auxiliary machinery; a paradise of mechanics. But when I went to sea for six months as part of the training, I was on a Victory ship with a sealed turbine and electrically driven auxiliary machinery. Very boring. Therefore, when the war ended with the atom bomb, I left the merchant marine and went to work for my father while waiting to return to college. I knew so little about physics that I didn't know even vaguely why the bomb was so powerful.

I didn't get right back to college. The draft was still in force in the United States. I was drafted, and spent a pleasant year at an army installation in Washington, DC, doing very little. Finally, I returned to the Polytechnic Institute and received a summa cum laude bachelor degree in Chemical Engineering in 1948.

The skills and knowledge I acquired at the Polytechlnic Institute have been crucial in all my experimental work: the use of strength of materials principles in equipment design, machine shop practice, engineering drawing, practical fluid mechanics, inorganic and organic chemistry, chemical laboratory techniques, manufacturing processes, metallurgy, basic concepts in mechanical engineering, basic concepts in electrical engineering, dimensional analysis, speed and power in mental arithmetic, numerical estimation (crucial when depending on a slide rule for calculations), and much more.

Upon graduation, I joined the General Electric Company. After a year in an advanced engineering training program, I settled in Schenectady, New

York, working as a Chemical Engineer in the Electron Tube Division. I worked in an engineering office in the electron tube production factory. Our job was to troubleshoot production problems, to improve production processes, and occasionally to do a little development work. We were not a fancy R&D office. I worked on speeding-up the production of television picture tubes, and then on problems of grid emission in industrial power tubes. These tasks led to a turning point in my life.

I had to learn a little about how electron vacuum tubes worked, so I took a few courses in Union College in Schenectady specifically, atomic physics and advanced calculus. I got to know a wonderful physics professor, Vladimir Rojansky. One day he said to me "Martin, what you are interested in is called physics not chemistry!" At the age of 23, I finally decided to begin the study of physics.

Graduate Study in Physics, I. I. Rabi, and Learning the Physicist's Trade

I entered the physics doctoral program in Columbia University in the autumn of 1950. Looking back, it seems amazing that I was admitted. True, I had a summa cum laude bachelor degree, but I had taken only two courses in physics: one year of elementary physics and a half-year of atomic physics. There were several reasons I could do this 1950; it could not have been done today. First, graduate study in physics was primitive in 1950, compared to today's standards. We did not study quantum mechanics until the second year, the first year was devoted completely to classical physics. The most advanced quantum mechanics we ever studied was a little bit in Heitler, and we were not expected to be able to do calculations in quantum electrodynamics.

Second, there was no thought of advising or course guidance by the Columbia Physics Department faculty–students were on their own. I was arrogant about my ability to learn anything fast. By the time I realized I was in trouble, but the time I realized that many of my fellow students were smarter than me and better trained then me, it was too late to quit. I had explained the return to school to my astonished parents by telling them that physics was what Einstein did. They thought if Einstein, why not Martin; I could not quit. I survived the Columbia Physics Department, never the best student, but an ambitious and hard-working student. I was married and had one child. I had to get my Ph.D and once more earn a living.

Just as the Polytechnic Institute was crucial in my learning how to do engineering; just as Union College and Vladimir Rojansky were crucial in my choosing physics; so Columbia University and my thesis advisor, I. I. Rabi, were crucial in my learning how to do experimental physics. I untertook for my doctoral research the problem of using the atomic beam resonance method to measure the quadrupole moment of the sodium nucleus. This measurement had to be made using an excited atomic state, and Rabi had found a way to do this.

As is well known, Rabi was not a "hands-on" experimenter. He never used tools or manipulated the apparatus. I learned experimental techniques from older gratuate students and by occasionally going to ask for help or advice from Rabi's colleague, Polykarp Kusch. I hated to go to Kusch, because it was always an unpleasant experience. He had a loud voice which he deliberately made louder so that the entire floor of students could hear about the stupid question asked by a graduate student.

Thus as in the course work, I was on my own in learning the experimenter's trade. I learned quickly, as I tell my graduate students now, there are no answers in the back of the book when the equipment doesn't work or the measurements look strange.

I learned things more precious than experimental techniques from Rabi. I learned the deep importance of choosing one's own research problems. Rabi once told me that he would worry when talking to Leo Szilard that Szilard would propose some idea to Rabi. This was because Rabi wouldn't carry out an idea suggested by someone else, even though he had already been thinking about that same idea.

I learned from Rabi the importance of getting the right answer and checking it thoroughly. When I finished my measurement of the quadrupole moment, I was eager to publish and to get on with earning a living. But Rabi had heard that a similar measurement had been made by an optical resonance method in France. He wrote to the French physicists to see if they had a similar answer. He didn't telephone or cable; he calmly wrote. I waited nervously. Six or eight weeks later he received the answer that they had a similar answer; then, I was allowed to publish. It is far better to be delayed, it is better to be second in publishing a result, than to publish first with the wrong answer.

It was Rabi who always emphasized the importance of working on a fundamental problem, and it was Rabi who sent me into elementary particle physics. It would have been natural for me to continue in atomic physics, but he preached particle physics to me–particulary when his colleagues in atomic physics were in the room. I think that most of that public preaching may have been Rabi's way of deliberately irritating his colleagues.

Michigan, Bubble Chambers, and On my Own with L.W. Jones

When I received my Ph.D. in 1955, I had job offers from the Physics Departments at Yale, the University of Illinois, and the University of Michigan. At that time, the first two Physics Departments had better reputations in elementary particle physics, and so I deliberately went to Michigan. I followed a two-part theorem that I always pass on to my graduate students and post doctoral research associates. Part 1: don't choose the most powerful experimental group or department–choose the group or department where you will have the most freedom. Part 2: there is an advantage in working in a small or new group–then you will get the credit for what you accomplish.

At Michigan I first worked in bubble chamber physics with Donald Glaser. But I wanted to be on my own. When the Russians flew SPUTNIK in 1957, I saw the opportunity, and jointly with my colleague, Lawrence W. Jones, we wrote to Washington for research money. We began our own research program, using first the now-forgotten luminescent chamber and then spark chambers. This brings me to the story I tell in my Nobel lecture on the discovery of the tau lepton.

It was Good Fortune...

Looking back to to my early years in Brooklyn, at the Polytechnic Institute, and at the General Electric Company, I am astonished to be writing a biographical note as a Nobel Laureate. I have tried to tell how it happened, yet I realize that I have left out the most crucial element: good fortune. It was good fortune to be a child during the Depression years and a youth during the war years. I lived in a country united by the belief that hard work and perseverance could get one through great difficulties. I saw right triumph. The progression of my career coincided with the growth of universities and the tremendous expansion in federal support for basic research, Academic jobs were relatively easy to get and hold, research funds were relatively easy to get. All good fortune. Of course, my ultimate good fortune was that the tau existed.

Life is much harder for the young women and men who are in science in present times. But they are smarter and better trained than I was at their ages; they know more and have better equipment. I wish them good fortune.

REFLECTIONS ON THE DISCOVERY OF THE TAU LEPTON

Nobel Lecture, December 8, 1995

by

MARTIN L. PERL

Stanford Linear Accelerator Center, Stanford University, Stanford, California 94309, USA

First thoughts

My first thoughts in writing this lecture are about the young women and young men who are beginning their lives in science: students and those beginning scientific research. I have been in experimental scientific research for 45 years; I have done some good experiments of which the best was the discovery of the tau lepton; I have followed research directions that turned out to be uninteresting; I have worked on experiments that failed. And so while recounting the discovery of the tau for which I have received this great honor, I will try to pass on what I have learned about doing experimental science.

I begin my reflections by going back in time before the tau, before even my interest in leptons. I was trained as an engineer at Polytechnic University (then the Polytechnic Institute of Brooklyn) and I always begin the design of an experiment with engineering drawings, with engineering calculations on how the apparatus is to be built and how it should work. My strong interest in engineering and in a mechanical view of nature carried over into my career in physics.

My doctoral thesis research (Perl, Rabi, and Senitzky 1955) was carried out at Columbia University in the early 1950's under Professor Isidor Rabi. The research used the atomic beam resonance method invented by Rabi, for which he received a Nobel Prize in 1944. My experimental apparatus, Fig. 1, was boldly mechanical with a brass vacuum chamber, a physical beam of sodium atoms, submarine storage batteries to power the magnets–and in the beginning of the experiment, a wall galvanometer to measure the beam current. I developed much of my style in experimental science in the course of this thesis experiment. When designing the experiment and when thinking about the physics, the mechanical view is always dominant in my mind. More important, my thinking about elementary particles is physical and mechanical. In the basic production process for tau leptons

$$e^+ + e^- \rightarrow \tau^+ + \tau^-. \tag{1}$$

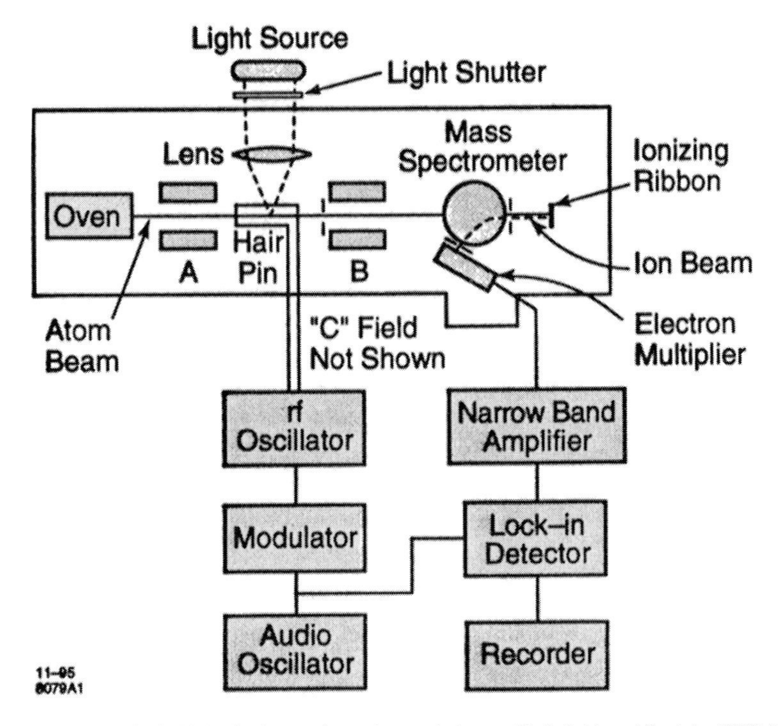

Figure 1. From the author's Ph.D. thesis experiment in atomic beams (Perl, Rabi, and Senitzky, 1955) The caption read:

"Schematic drawing of the apparatus. The light source is shown on the side of the apparatus for clarity, but it actually lies above the apparatus. The C magnet, which produces a homogeneous field in the "hair-pin," is also not shown for clarity. The six external boxes which represent the major electronic components do not indicate the physical position of the components."

I see the positron, e^+, and electron, e^-, as tiny particles which collide and annihilate one another. I see a tiny cloud of energy formed which we technically call a virtual photon, $\gamma_{virtual}$; and then I see that energy cloud change into two tiny particles of new matter–a positive tau lepton, τ^+, and a negative tau lepton, τ^-.

In my thesis experiment I first experienced the pleasures, the anxieties, and sometimes the pain, that is inherent in experimental work: The pleasure when an experiment is completed and the data safely recorded, the anxiety when an experiment does not work well or breaks, the pain when an experiment fails or when an experimenter does something stupid. In my thesis experiment the acquisition of a set of data took about a day, and so there were several alternating periods of anxiety and pleasure within a week. When I broke a McCloud vacuum gauge and spread mercury inside the vacuum chamber, the pain of restoring the apparatus lasted but a few weeks. At the other extreme, in the discovery of the tau the ups and downs of my emotions extended over years. This brings me to the research which led me to think about heavy leptons.

From strong interactions to the electron-muon problem

In eight wonderful and productive years at the University of Michigan, I learned the experimental techniques of research in elementary particle physics (scintillation counters, bubble chamber, trigger electronics, and data analysis) working with my research companions, Lawrence Jones, Donald Meyer, and later Michael Longo. We learned these techniques together, often adding our own new developments. One of the most pleasurable experiences was the development of the luminescent chamber, Fig. 2, by Jones and me with the help of our student Kwan Lai (Lai, Jones, and Perl 1961). We photographed and recorded the tracks of charged particles in a sodium iodide crystal using primitive electron tubes which intensified the light coming from the track.

I worked in the physics of strong interactions. Jones and I, using spark chambers, carried out at the Bevatron a neat set of measurements on the elastic scattering of pions on protons (Damouth, Jones, and Perl 1963; Perl, Jones, and Ting 1963). Later, after I left the University of Michigan for Stanford University, Longo and I, working with my student Michael Kreisler, initiated a novel way to measure the elastic scattering of neutrons on protons (Kreisler et al. 1966).

These elastic scattering experiments pleased me in many ways. The equipment was bold and mechanical, with large flashing spark chambers and a camera with a special mechanism for quick movement of the film. Data acquisition was fast, and the final data was easily summarized in a few graphs, but I gradually became dissatisfied with the theory needed to explain our measurement. I am a competent mathematician but I dislike complex mathematical explanations and theories, and in the 1950's and 1960's the theory of strong interactions was a complex mess, going nowhere. I began to think about the electron and the muon, elementary particles which do not partake in the strong interaction.

The electron was discovered in the late nineteenth century; the final characterization of its nature was achieved by J. J. Thomson in the 1890's. He received a Nobel Prize in 1906 for investigation of electrical conduction in gases. The muon was found in cosmic rays in the 1930's. Table 1 lists their properties as known in the 1960's; this table is still correct today

Table 1. Properties of the electron and muon. The electric charge is given in units of 1.60×10^{-19} coulombs. The mass is given in units of the mass of the electron 9.11×10^{-31} kilograms.

Particle	**Electron**	**Muon**
Symbol	e	μ
Electric charge	+1 or −1	+1 or −1
Mass	1	206.8
Does particle have electromagnetic interactions?	yes	yes
Does particle have weak interactions?	yes	yes
Does particle have strong interactions?	no	no
Associated neutrino	ν_e	ν_μ
Associated antineutrino	$\bar{\nu}_e$	$\bar{\nu}_\mu$
Lifetime	Stable	2.2×10^{-6} sec

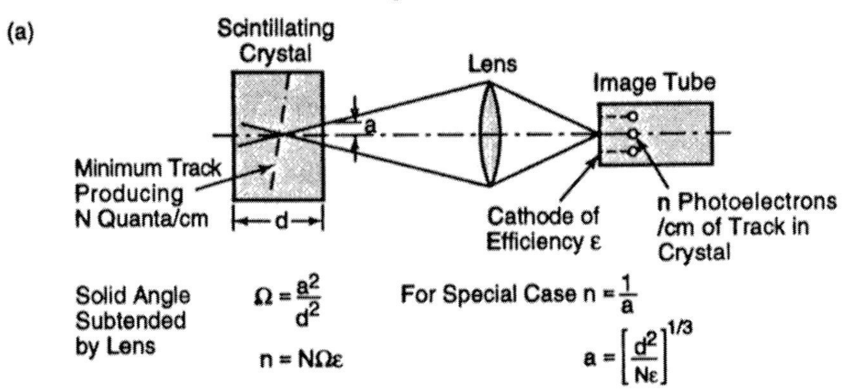

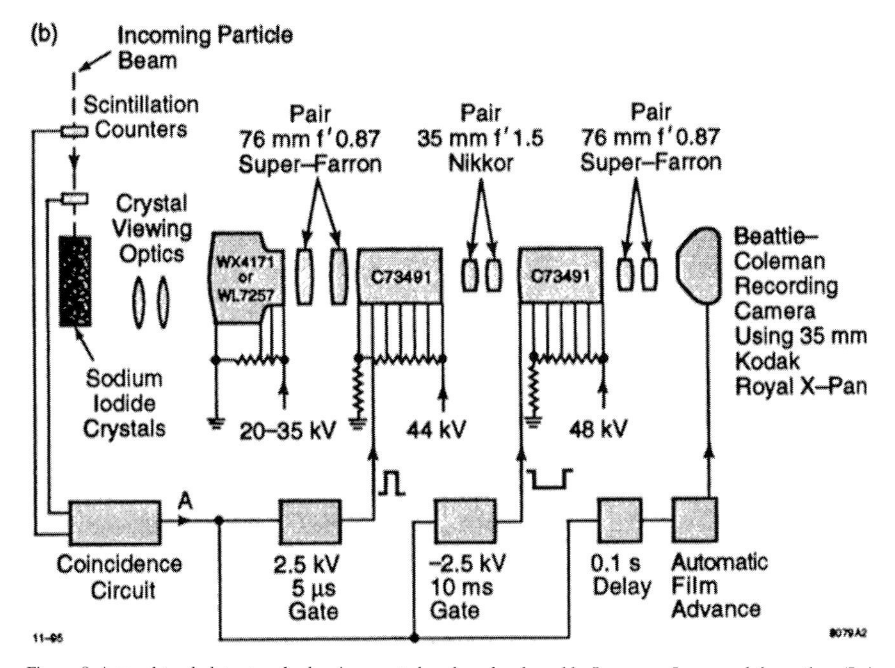

Figure 2. A novel track detector, the luminescent chamber, developed by Lawrence Jones and the author (Lai, Jones, and Perl, 1961) before the advent of the optical spark chamber. The caption read:

"(a) Relationships between track resolution, a, depth of field, d, and track information, n, for the homogeneous luminescent chamber. For NaI(T1) in which we have $N=10^5$, n=1.7 mm for d=10 cm and n=10 photoelectrons per cm of track in the crystal.

(b) Schematic diagram of the luminescent chamber system currently in use. The chamber-viewing optics and beam-defining scintillation counters are oversimplified and generalized in this diagram."

There were two puzzles about the relation between the electron and the muon. First, as shown in the table, the properties with respect to particle interactions are the same for the electron and the muon, but the muon is 206.8 times heavier. Why? The second puzzle is that since the muon is unstable, with an average lifetime of 2.2×10^{-6} seconds decay to an electron, one expects that the decay process would be

$$\mu^- \rightarrow e^- + \gamma$$
$$\mu^+ \rightarrow e^+ + \gamma. \tag{2}$$

Here γ means a photon, and the expectation would be that the γ carries off the excess energy produced by the difference between the muon mass and the electron mass–but this is not the nature of the muon or the electron. The muon decays to an electron by a complicated process,

$$\mu^- \rightarrow e^- + \overline{\nu}_e + \nu_\mu$$
$$\mu^+ \rightarrow e^+ + \nu_e + \overline{\nu}_\mu, \tag{3}$$

in which a neutrino and an antineutrino are produced. There is something in the nature of the muon which is different from the nature of the electron. By the late 1950's, there was the electron-muon problem (e–μ problem) with two parts:

* Why is the muon 206.8 times heavier than the electron?
* Why doesn't the muon decay through the process $\mu \rightarrow e + \gamma$

While I was at the University of Michigan, I was intrigued by the careful measurements being made on the (g-2) of the muon by Charpak et al. (1962) at CERN, and on the (g-2) of the electron by Wilkinson and Crane (1963) at Michigan. I was also interested in the precision studies of positronium and muonium then in progress, as well as other precision atomic physics experiments. These low energy studies of the charged leptons were in very capable hands, and I could not see how I could contribute.

I knew about the pioneer low energy, neutrino experiments of Frederick Reines and Clyde Cowan, Jr. I must interrupt my narrative to quote two momentous sentences from Reines and Cowan (1953):

> "An experiment has been performed to detect the free neutrino. It appears probable that this aim has been accomplished although further confirmatory work is in progress."

These were extraordinarily difficult experiments, and again I could not see how I could contribute.

I am honored to share this year's Nobel Prize in Physics with Frederick Reines, and I am sad that Clyde Cowan, Jr. is not alive to share this honor.

As for high energy neutrino experiments, they were already being carried out by the powerful set of Nobel Laureates, Leon Lederman, Melvin Schwartz and Jack Steinberger (Danby et al. 1962).

I reflected that it would be most useful for me to consider high-energy experiments on charged leptons, experiments which might clarify the nature of the lepton or explain the electron-muon problem. This is a research strategy that I have followed quite a few times in my life. I stay away from lines of research where many people are working, and in particular I stay away from lines of research where very smart and competent people are working. I find it more comfortable to work in uncrowded areas of physics.

I caution the young scientist about this advice. Almost all the time the best experimenters and the most experimenters work in the most fruitful area. If there are few or no investigators working on a problem, it may be an unproductive problem. In the end, it is a question of temperament and comfort.

SLAC, leptons, and heavy leptons

In 1962, the opportunity arose to think seriously about high-energy experiments on charged leptons when Wolfgang K. H. Panofsky and Joseph Ballam offered me a position at the yet-to-be-built Stanford Linear Accelerator Center (SLAC). Here was a laboratory which would have primary electron beams; a laboratory at which one could easily obtain a good muon beam; a laboratory in which one could easily obtain a good photon beam for production of lepton pairs. And on the Stanford campus at the High Energy Physics Laboratory, the Princeton–Stanford $e^- e^-$ storage ring was operating (O'Neil et al. 1958, Barber et al. 1966).

When I arrived at SLAC in 1963, I began to plan various attacks on, and investigations of, the electron–muon problem. Although the linear accelerator would not begin operation until 1966, my colleagues and I began to design and build experimental equipment. The proposed attacks and investigations were of two classes. In one class, I proposed to look for unknown differences between the electron and the muon; the only known differences being the mass difference and the observation that the decay reaction $\mu \rightarrow e + \gamma$ does not occur. The other class of proposed attacks and investigations was based on my speculation that there might be more leptons similar to the electron and the muon, unknown heavier charged leptons. I dreamed that if one could find a new lepton, the properties of the new lepton might teach us the secret of the electron-muon puzzle.

My first attack used an obvious idea. An intense photon (γ) beam could be made at SLAC using the reactions.

$$e^- + \text{nucleus} \rightarrow \gamma + \ldots . \quad (4)$$

The photons so produced could then interact with another nucleus to produce a pair of charged particles , x^+ and x^-,

$$\gamma + \text{nucleus} \rightarrow x^+ + x^- + \ldots . \quad (5)$$

Any pair of charged particles could be produced if the γ had enough energy. To the young experimenter, I remark that there is nothing wrong with an obvious experimental idea as long as you are the first to use the idea.

My hope was that we would find a new x particle, perhaps a new charged lepton somehow related to the electron or muon. A vague hope by the standards of our knowledge of elementary particle physics today. We were certainly naive in the 1960's.

We didn't find any new leptons or any new particles of any kind (Barna et al. 1968); as we now know, there were no new particles to find given the experimental limitations of this search experiment. The search used the pair-production calculations of Tsai and Whitis (1966); this experiment was the beginning of a long and fruitful collaboration between my colleague Y.-S (Paul) Tsai and myself.

Studies of muon-proton inelastic scattering

Although this first attempt to penetrate the mysteries of the electron and muon failed, we were already preparing to study muon-proton inelastic scattering

$$\mu^- + p \rightarrow \mu^- + \text{hadrons}$$

to compare it with electron - proton inelastic scattering,

$$e^- + p \rightarrow e^- + \text{hadrons} \ ,$$

Extensive studies of e–p inelastic scattering were planned at SLAC. Indeed, some of those studies led to the 1990 Nobel Physics Prize being awarded to Jerome Friedman, Henry Kendall, and Richard Taylor. My hope was that we would find a difference between the μ and e other than the differences of mass and lepton number. In particular, I hoped that we would find a difference at large momentum transfers–another naive hope when viewed by our knowledge today of particle physics. For example, I speculated (Perl 1971) that the muon might have a special interaction with hadrons not possessed by the electron, see Fig. 3.

It is always a good plan for a speculative experimenter to have two experiments going, or at least one going and one being built. Of course, that was easier in the 1960's than it is now, since most modern high-energy physics

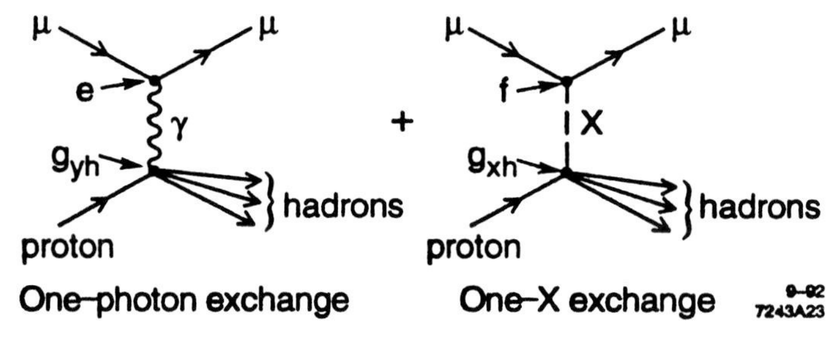

Figure 3. From Perl (1971): "The interaction of a muon with hadrons through exchange of a particle X, an example of the speculation that the muon has a special interaction with hadrons that is not possessed by the electron."

experiments are so large and complicated. Still, it can be done. My main present research is in tau physics, working with the CLEO Collaboration which uses the 10 GeV electron-positron collider CESR at Cornell University. But there is also a small nonaccelerator experiment at SLAC in which a few graduate students (Eric Lee, Nancy Mar, Manuel Ortega), a few colleagues, and myself are searching for free quarks.

Returning to the late 1960's, my colleagues and myself measured the differential cross sections for inelastic scattering of muons on protons, and then compared the μ–p cross sections with the corresponding e–p cross sections (Toner et al. 1972, Braunstein et al. 1972). We were looking for a difference in magnitude, or a difference in behavior of the cross sections. As discussed in Perl and Rapidis (1974), these differences could come from a new nonelectromagnetic interaction between the μ and hadrons or from the μ not being a point particle. However as summarized in Toner et al. (1972), we found no significant deviation.

Other experimenters studied the differential cross section for μ–p elastic scattering and compared it with e–p elastic scattering (Ellsworth et al. 1960, Camilleri et al. 1969, Kostoulas et al. 1974), but statistically significant differences between μ–p and e–p cross sections could not be found in either the elastic or inelastic case. Furthermore, there were systematic errors of the order of 5 or 10% in comparing μ–p and e–p cross sections because the techniques used were so different.

Experimental science is a craft and an art, and part of the art is knowing when to end a fruitless experiment. There is a danger of becoming obsessed with an experiment even if it goes nowhere. I avoided obsession and gave up. That turned out to be a good decision because modern experiment has shown that the scattering experiment does not illuminate any differences between the electron and the muon beyond the mass difference.

Heavy leptons in the 1960's

While building the apparatus using our muon-proton inelastic scattering experiment, and during the first operation of that experiment, I was thinking of another way to look for new charged leptons, L, using the reaction,

$$e^+ + e^- \to L^+ + L^- .$$

Before turning to this third attack on the electron-muon problem, I describe the general thinking in the 1960's about the possible existence and types of new leptons. By the beginning of the 1960's, there were papers on the possibility of the existence of charged leptons more massive than the e and μ. I remember reading the 1963–1964 papers of Zel'dovich (1963), of Lipmanov (1964), and of Okun (1965). Since the particle generation concept was not yet an axiom of our field, older models of particle relationships were used. For example, if one thought (Low 1965) that there might be an electromagnetic excited state e* of the e, then the proper search method was

$$e^- + \text{nucleon} \rightarrow e^{-*} + \ldots ,$$
$$e^{-*} \rightarrow e^- + \gamma . \tag{6}$$

If one thought (Lipmanov 1964) that there was a μ' which was a member of a μ, ν_μ, μ' family, then the proper search method was

$$\nu_\mu + \text{nucleon} \rightarrow \mu^{-\prime} + \ldots . \tag{7}$$

It is interesting to note, in view of the search a decade later for $\tau^- \rightarrow \nu_\tau \pi^-$, that Lipmanov (1964) calculated the branching fraction for this decay mode.

By the second half of the 1960's, the concept had been developed of a heavy lepton L and its neutrino ν_L forming an L, ν_L pair. Thus, in a paper written in 1968, Rothe and Wolsky (1969) discuss the lower mass limit on such a lepton set by its absence in *K* decays. They also discuss the decay of such a lepton into the modes

$$L \rightarrow e\, \bar{\nu}_e\, \nu_L\, ,\ \mu \bar{\nu}_\mu\, \nu_L\, ,\ \pi \nu_L\, .$$

Electron–positron colliding beams and sequential leptons

The construction and operation of electron-positron colliders began in the 1960's (Voss 1994). By September 1967 at the Sixth International Conference on High Energy Accelerators, Howard (1967) was able to list quite a few electron-positron colliders. There was the pioneer 500 MeV ADA collider already operated at Frascati in the early 1960's and, also at Frascati, ADONE was under construction. The 1 GeV ACO at Orsay and 1.4 GeV VEPP–2 at Novosibirsk were in operation. The 6 GeV CEA Collider at Cambridge was being tested, and colliders had been proposed at DESY and SLAC (Ritson et al. 1964).

The 1964 SLAC proposal (Ritson et al. 1964), see Fig. 4, had already discussed the reaction

$$e^+ + e^- \rightarrow x^+ + x^- . \tag{8}$$

Of course, x might be a charged lepton. This proposal did not directly lead to the construction of an e^+e^- collider at SLAC because we could not get the funding. About five years later–with the steadfast support of the SLAC director, Wolfgang Panofsky, and with a design and construction team led by Burton Richter–construction of the SPEAR e^+e^- collider was begun at SLAC.

It was this 1964 proposal and the 1961 seminal paper of Cabibbo and Gatto (1961) entitled, "Electron-Positron Colliding Beam Experiments," which focused my thinking on new charged lepton searches using an e^+e^- collider. As we carried out the experiments described previously, I kept loo-

PROPOSAL FOR A HIGH-ENERGY
ELECTRON-POSITION COLLIDING-BEAM STORAGE RING
AT THE
STANFORD LINEAR ACCELERATOR

March 1964

It is proposed that the Atomic Energy Commission support the construction at Stanford University of a Colliding-Beam Facility (storage ring) for high-energy electrons and positrons. This facility would be located at the Stanford Linear Accelerator Center, and it would make use of the SLAC accelerator as an injector.

This proposal was prepared by the following persons:

Stanford Physics Department

D. Ritson

Stanford linear Accelerator Center

S. Berman
A. Boyarski
F. Bulos
E. L. Garwin
W. Kirk
B. Richter
M. Sands

Figure 4. The cover page of the 1964 SLAC proposal to build an electron-positron collider (D. Ritson et al. 1964).

king for a model for new leptons–a model which would lead to definitive colliding beam searches, while remaining reasonably general. Helped by discussions with my colleagues, such as Paul Tsai and Gary Feldman, I came to what I later called the sequential lepton model.

I thought of a sequence of pairs

$$\begin{array}{ll} e^- & \nu_e \\ \mu^- & \nu_\mu \\ L^- & \nu_L \\ L'^- & \nu_{L'} \end{array} \tag{9}$$

each pair having a unique lepton number. I usually thought about the leptons as being point Dirac particles. Of course, the assumptions of unique lepton number and point particle nature were not crucial, but I liked the simplicity. After all, I had turned to lepton physics in the early 1960's in a search for simple physics.

The idea was to look for

$$e^+ + e^- \to L^+ + L^- , \tag{10a}$$

with

$$\begin{aligned} L^+ &\to e^+ + \text{undetected neutrinos carrying off energy} \\ L^- &\to \mu^- + \text{undetected neutrinos carrying off energy} , \end{aligned} \tag{10b}$$

or

$$\begin{aligned} L^+ &\to \mu^+ + \text{undetected neutrinos carrying off energy} \\ L^- &\to e^- + \text{undetected neutrinos carrying off energy} . \end{aligned} \tag{10c}$$

This search method had many attractive features:

- If the L was a point particle, we could search up to an L mass almost equal to the beam energy, if we had enough luminosity.
- The appearance of an $e^+\mu^-$ or $e^-\mu^+$ event with missing energy would be dramatic.
- The apparatus we proposed to use to detect the reactions in Eqs. 10 would be very poor in identifying types of charged particles (certainly by today's standards) but the easiest particles to identify were the e and the μ.
- There was little theory involved in predicting that the L would have the weak decays

$$\begin{aligned} L^- &\to \nu_L + e^- + \bar{\nu}_e \\ L^- &\to \nu_L + \mu^- + \bar{\nu}_\mu , \end{aligned} \tag{11}$$

  with corresponding decays for the L^+. One simply could argue by analogy from the known decay

$$\mu^- \to \nu_\mu + e^- + \bar{\nu}_e .$$

I incorporated the search method summarized by Eqs. 10 in our 1971 Mark I proposal to use the not-yet-completed SPEAR e^+e^- storage ring.

My thinking about sequential leptons and the use of the method of Eqs. 10 to search for them was greatly helped and influenced by two seminal papers of Paul Tsai. In 1965, he published with Anthony Hearn the paper, "Differential Cross Section for $e^+ + e^- \to W^+ + W^- \to e^- + \bar{\nu}_e + \mu^+ + \nu_\mu$," (Tsai and Hearn 1965).

This work discussed finding vector boson pairs W^+W^- by their eμ decay mode. It was thus closely related to my thinking, described above, of finding L^+L^- pairs by their eμ decay mode. Tsai's 1971 paper entitled, "Decay Correlations of Heavy Leptons in $e^+ + e^- \rightarrow L^+ + L^-$," provided the detailed theory for the applications of the sequential lepton model to our actual searches (Tsai 1971). Thacker and Sakurai (1971) also published a paper on the theory of sequential lepton decays, but it is not as comprehensive as the work of Tsai. Also important to me was the general paper, "Spontaneously Broken Gauge Theories of Weak Interactions and Heavy Lepton," by James Bjorken and Chris Llewellyn-Smith (1973).

The SLAC–LBL proposal

After numerous funding delays, a group led by Burton Richter and John Rees of SLAC Group C began to build the SPEAR e^+e^- collider at the end of the 1960's. Gary Feldman and I, and our Group E, joined with their Group C and a Lawrence Berkeley Laboratory Group led by William Chinowsky, Gerson Goldhaber, and George Trilling to build the Mark I detector. In 1971, we submitted the SLAC-LBL Proposal (Larsen et al. 1971) using the Mark I detector at SPEAR. (The detector was originally called the SLAC–LBL detector and only called the Mark I detector when we began to build the Mark II detector. For the sake of simplicity, I refer to it as the Mark I detector.) The contents of the proposal consisted of five sections and a supplement, as follows:

A.	Introduction	Page 1
B.	Boson Form Factors	Page 2
C.	Baryon Form Factors	Page 6
D.	Inelastic Reactions	Page 12
E.	Search for Heavy Leptons	Page 16
	Figure Captions	Page 19
	References	Page 20
	Supplement	

The heavy lepton search was left for last, and allotted just three pages because to most others it seemed a remote dream. But the three pages did contain the essential idea of searching for heavy leptons using eμ events, Eqs. 10.

I wanted to include a lot more about heavy leptons and the e–μ problem, but my colleagues thought that would unbalance the proposal. We compromised on a 10-page supplement entitled, "Supplement to Proposal SP-2 on Searches for Heavy Leptons and Anomalous Lepton-Hadron Interactions." The supplement began as follows:

"While the detector is being used to study hadronic production processes it is possible to simultaneously collect data relevant to the following questions:

(1) Are there charged leptons with masses greater than that of the muon?

We normally think of the charged heavy leptons as having spin 1/2 but the search method is not sensitive to the spin of the particle. This search for charged heavy leptons automatically includes a search for the intermediate vector boson which has been postulated to explain the weak interactions.

(2) Are there anomalous interactions between the charged leptons and the hadrons?

In this part of the proposal we show that using the detector we can gather definitive information on the first question within the available mass range. We can obtain preliminary information on the second question–information which will be very valuable in designing further experiments relative to that question. We can gather all this information while the detector is being used to study hadronic production processes. Additional running will be requested if the existence of a heavy lepton, found in this search, needs to be confirmed."

My first interest was to look for heavy leptons, but I still had my old interest of looking for an anomalous lepton interaction, the idea that led to the study of muon-proton inelastic scattering.

Lepton searches at ADONE

While SPEAR and the Mark I detector were being built, lepton searches were being carried out at the ADONE e^+e^- storage ring by two groups of experimenters in electron-positron annihilation physics: One group reported in 1970 and 1973 (Alles-Borelli et al. 1970, Bernardini et al. 1973). In the later paper, they searched up to a mass of about 1 GeV for a conventional heavy lepton and up to about 1.4 GeV for a heavy lepton with decays restricted to leptonic modes. The other group of experimenters in electron-positron annihilation physics was led by Shuji Orito and Marcello Conversi. Their search region (Orito et al. 1974) also extended to masses of about 1 GeV.

Discovery of the tau in the Mark I experiment: 1974–1976

SPEAR and the Mark I Detector

The SPEAR e^+e^- collider began operation in 1973. Eventually SPEAR obtained a total energy of about 8 GeV; but in the first few years, the maximum energy with useful luminosity was 4.8 GeV. We began operating the Mark I experiment in 1973 in the form shown in Fig. 5. The Mark I was one of the first large-solid-angle, general purpose detectors built for colliding beams.

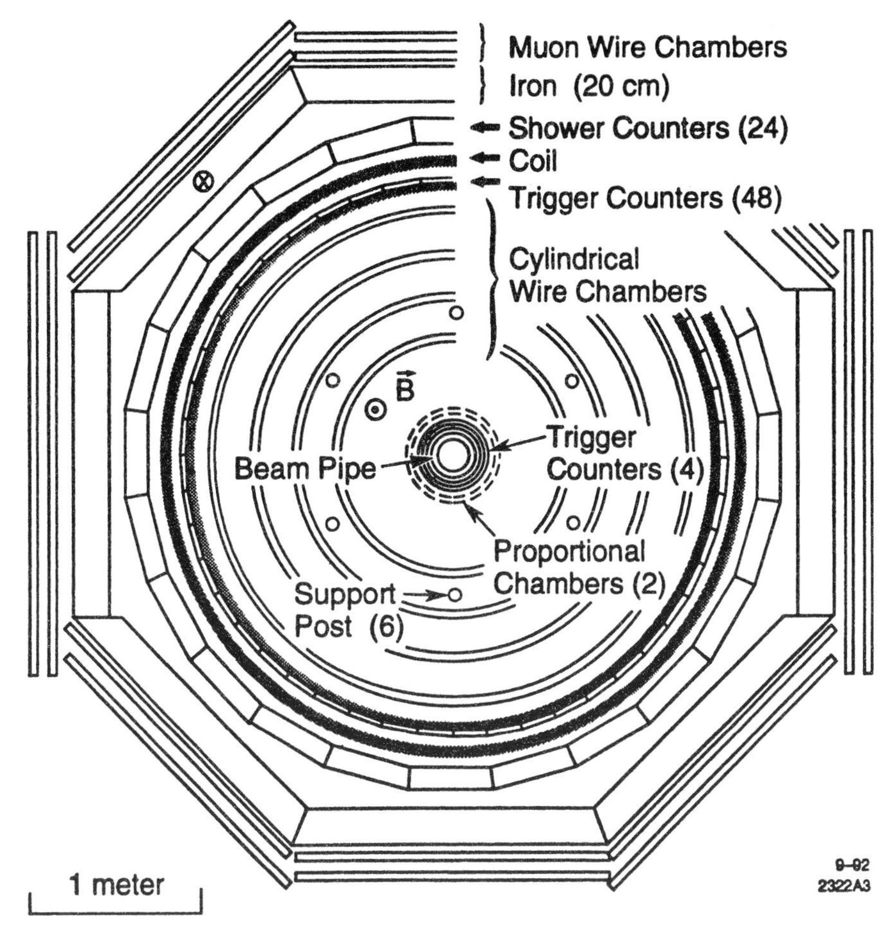

Figure 5. The initial form of the Mark I detector.

The use of large-solid-angle particle tracking and the use of large-solid-angle particle identification systems is obvious now, but it was not obvious twenty years ago. The electron detection system used lead-scintillator sandwich counters built by our Berkeley colleagues. The muon detection system was also crude, using the iron flux return which was only 1.7 absorption lengths thick.

The 1975 Canadian talks

In June 1975, I gave my first international talk on the e–μ events (Perl 1975a) at the 1975 Summer School of the Canadian Institute for Particle Physics. This was the second of my two lectures on electron-positron annihilation at the School.

The contents of the 1975 Summer School talk are shown below:

Contents of the 1975 Summer School talk

1. Introduction
   - A. Heavy Leptons
   - B. Heavy Mesons
   - C. Intermediate Boson
   - D. Other Elementary Bosons
   - E. Other Interpretations
2. Experimental Method
3. Search Method and Event Selection
   - A. The 4.8 GeV Sample
   - B. Event Selection
4. Backgrounds
   - A. External Determination
   - B. Internal Determination
5. Properties of eμ Events
6. Cross Sections of eμ Events
7. Hypothesis Tests and Remarks
   - A. Momentum Spectra
   - B. q_{coll} Distribution
   - C. Cross Sections and Decay Ratios
8. Compatibility of e^+e^- and $\mu^+\mu^-$ Events
9. Conclusions

This talk had two purposes. First, to discuss possible sources of e–μ events: heavy leptons, heavy mesons, or intermediate bosons; second, to demonstrate that we had good evidence for e–μ events. The largest single energy data sample (Table 2) was at 4.8 GeV, the highest energy at which we could then run SPEAR. The 24 e–μ events in the total charge = 0, number photons = 0 column was our strongest claim.

One of the cornerstones of this claim was an informal analysis carried out by Jasper Kirkby, who was then at Stanford University and at SLAC. He showed me that just using the numbers in the 0 charge, 0 photons columns of Table 2, we could calculate the probabilities for hadron misidentification in this class of events. There were not enough eh, μh, and hh events to explain away the 24 e–μ events.

The misidentification probabilities determined from three-or-more prong hadronic events and other considerations are given in Table 3. Compared to present experimental techniques, the $P_{h \rightarrow e}$ and $P_{h \rightarrow \mu}$ misidentification probabilities of about 0.2 are enormous, but I could still show that the 24 e–μ events could not be explained away.

And so the evidence for a new phenomena was quite strong–not incon-

Table 2. From Perl (1975a). A table of 2-charged-particle events collected at 4.8 GeV in the Mark I detector. The table, containing 24 eμ events with zero total charge and no photons, was the strongest evidence at that time for the τ. The caption read:

"Distribution of 513, 4.8 GeV, 2-prong, events which meet the criteria: p_e >0.65 GeV/c, p_μ> 0.65 GeV/c, ϕ_{copl} >20°".

	Total Charge = 0			**Total Charge = ±2**		
Number photons =	0	1	>1	0	1	>1
ee	40	111	55	0	0	0
eμ	24	8	8	0	0	3
μμ	16	15	6	0	0	0
eh	18	23	32	2	3	3
μh	15	16	31	4	0	5
hh	13	11	30	10	4	6
Sum	126	184	162	16	8	17

Table 3. From Perl (1975a). The caption read:

"Misidentification probabilities for 4.8 GeV sample"

Momentum range (GeV/c)	$P_{h \to e}$	$P_{h \to \mu}$	$P_{h \to h}$
0.6– 0.9	.130±.005	.161±.006	.709±.012
0.9–1.2	.160±.009	.213±.011	.627±.020
1.2– 1.6	.206±.016	.216±.017	.578±.029
1.6– 2.4	.269±.031	.211±.027	.520±.043
Weighted average using hh, μh, and eμ events	.183±.007	.198±.007	.619±.012

trovertible, but still strong. What was the new phenomena: a sequential heavy lepton; a new heavy meson with the decays

$$M^- \to e^- + \bar{\nu}_e$$
$$M^- \to \mu^- + \bar{\nu}_\mu$$

My Canadian lecture ended with these conclusions:

"1) No conventional explanation for the signature e–μ events has been found.

2) The hypothesis that the signature e–μ events come from the production of a pair of new particles–each of mass about 2 GeV–fits almost all the data. Only the θ_{coll} distribution is somewhat puzzling.

3) The assumption that we are also detecting ee and μμ events coming from these new particles is still being tested."

I was still not able to specify the source of the eμ events: leptons, mesons or bosons. But I remember that I felt strongly that the source was heavy leptons. It would take two more years to prove that.

First publication: "We have no conventional explanation for these events"

As 1974 passed, we acquired e^+e^- annihilation data at more and more energies, and at each of these energies there was an anomalous e–μ event signal, see Fig. 6. Thus, I and my colleagues in the Mark I experiment became more and more convinced of the reality of the e–μ events and the absence of a conventional explanation. An important factor in this growing conviction was the addition of a special muon detection system to the detector (Fig. 7a), called the muon tower. This addition was conceived and built by Gary Feldman. Although we did not use events such as those in Fig. 7b in our first publication, seeing a few events like this was enormously comforting.

Finally, in December 1975, the Mark I experimenters published Perl et al. (1975b) entitled, "Evidence for Anomalous Lepton Production in $e^+ - e^-$ Annihilation."

The final paragraph reads:

"We conclude that the signature e–μ events cannot be explained either by the production and decay of any presently known particles or as coming

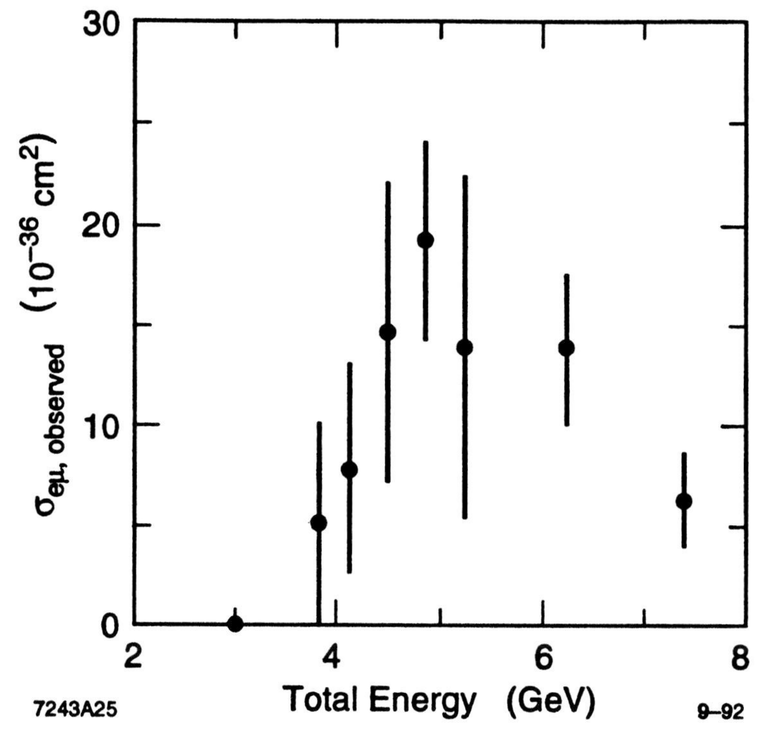

Figure 6. From Perl et al. (1975b): "The observed cross section for the signature eμ events from the Mark I experiment at SPEAR. This observed cross section is not corrected for acceptance. There are 86 events with a calculated background of 22 events."

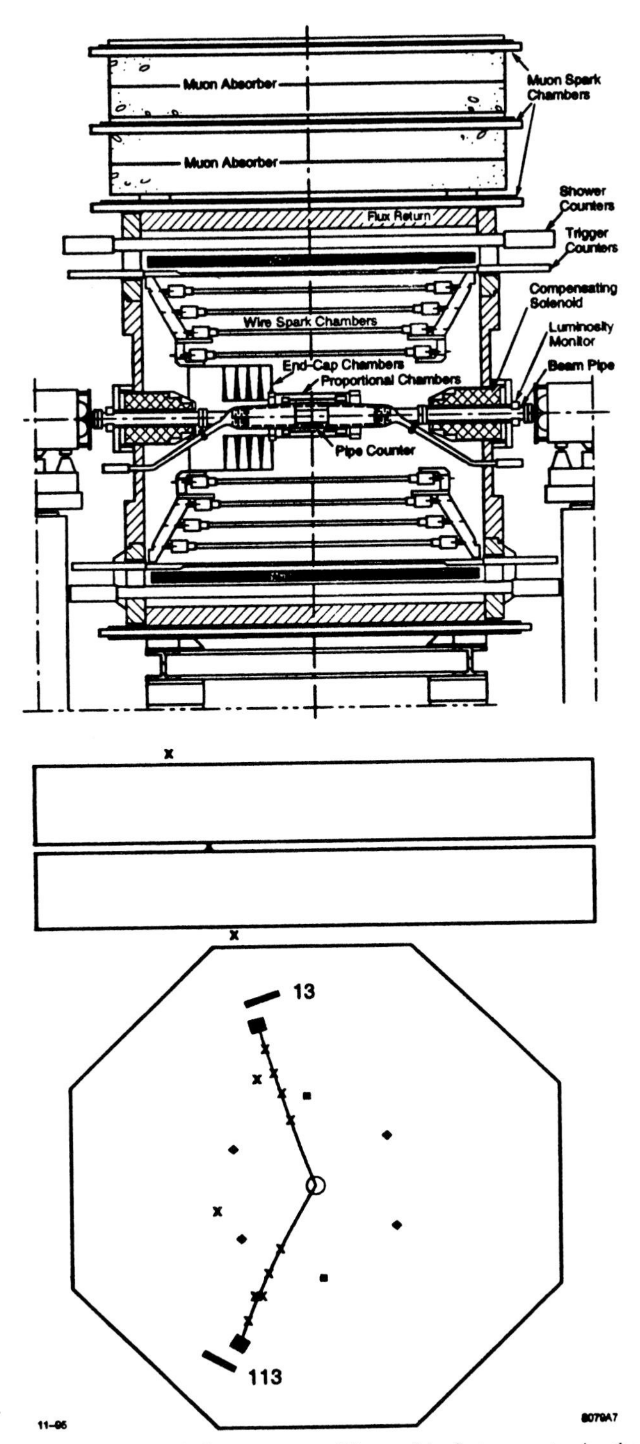

Figure 7. (a) The Mark I detector with the muon tower: (b) one of the first eμ events using the tower. The μ moves upward through the muon detector tower and the e moves downward. The numbers 13 and 113 give the relative amounts of electromagnetic shower energy deposited by the μ and e. The six square dots show the positions of longitudinal support posts of the magnetostrictive spark chamber used for tracking.

from any of the well-understood interactions which can conventionally lead to an e and a μ in the final state. A possible explanation for these events is the production and decay of a pair of new particles, each having a mass in the range of 1.6 to 2.0 GeV/c^2."

We were not yet prepared to claim that we had found a new charged lepton, but we were prepared to claim that we had found something new. To accentuate our uncertainty I denoted the new particle by "U" for unknown in some of our 1975–1977 papers. The name τ was suggested to me by Petros Rapidis, who was then a graduate student and worked with me in the early 1970's on the e–μ problem (Perl and Rapidis 1975). The letter τ is from the Greek triton for third–the third charged lepton.

Thus in 1975, twelve years after we began our lepton physics studies at SLAC, these studies finally bore fruit. But we still had to convince the world that the e–μ events were significant and we had to convince ourselves that the e–μ events came from the decay of a pair of heavy leptons.

Reflections on the discovery

This is a good place to reflect on the elements of the research which led to the discovery of the tau. First I had chosen a research area in which there were few investigators. Second, we had cast a wide net in studying the electron-muon problem: an attempt to photoproduce new leptons, experimental comparisons of muon-proton inelastic scattering with electron-proton inelastic scattering, and the use of the general reaction $e^+ + e^- \rightarrow L^+ + L^-$ to try to produce a heavy lepton. Third, a new technology, the electron-positron collider was available to carry out the L^+L^- production. Fourth, I had a good way to detect the L^+L^- production, namely the search for eμ events without photons. Fifth, I had smart. resourceful and patient research companions. I think these are representative of the elements which should be present in speculative experimental work; a broad general plan, specific research methods, new technology, and first-class research companions.Of course the element of luck will in the end be dominant. I had two great pieces of luck. First, there was a heavy lepton within the energy range of the SPEAR collider. Second, the Mark I experimental apparatus was sufficiently good to enable us to identify the e–μ events and prove their existence.

Is it a lepton? 1976–1978

Our first publication was followed by several years of confusion and uncertainty about the validity of our data and its interpretation. It is hard to explain this confusion a decade later when we know that τ pair production is 20% of the e^+e^- annihilation cross section below the Z°, when ττ pair events stand out so clearly at the Z°.

There were several reasons for the uncertainties of that period. It was hard to believe that both a new quark (charm) and a new lepton (tau) would be found in the same narrow range of energies. Also, while the existence of a fourth quark was required by theory, there was no such requirement for a third charged lepton, so there were claims that the other predicted decay modes of tau pairs, such as e–hadron and μ–hadron events, could not be found. Indeed, finding such events was just at the limit of the particle identification capability of the detectors of the mid-1970's.

Perhaps the greatest impediment to the acceptance of the τ as the third charged lepton was that there was no other evidence for a third particle generation. Two sets of particles–u, d, e^-, ν_e and c, s, μ^-, ν_μ–seemed acceptable, a kind of doubling of particles. But why three sets? A question which to this day has no answer.

It was a difficult time. Rumors kept arriving of definitive evidence against the τ: e–μ events not seen, the $\tau \to \pi\nu$ decay not seen, theoretical problems with momentum spectra or angular distribution. With colleagues such as Gary Feldman, I kept going over our data again and again. Had we gone wrong somewhere in our data analysis?

Clearly other tau pair decay modes had to be found. Assuming the τ to be a charged lepton with conventional weak interactions, simple and very general theory predicted the branching fractions

$$B(\tau^- \to \nu_\tau + e^- + \bar{\nu}_e) \approx 20\%$$

$$B(\tau^- \to \nu_\tau + \mu^- + \bar{\nu}_\mu) \approx 20\%$$

$$B(\tau^- \to \nu_\tau + \text{hadrons}) \approx 60\% \ .$$

Experimenters therefore should be able to find the decay sequences

$$\begin{aligned} &e^+ + e^- \to \tau^+ + \tau^- \\ &\tau^+ \to \bar{\nu}_\tau + \mu^+ + \nu_\mu \\ &\tau^- \to \nu_\tau + \text{hadrons}\ , \end{aligned} \tag{13}$$

and

$$\begin{aligned} &e^+ + e^- \to \tau^+ + \tau^- \\ &\tau^+ \to \bar{\nu}_\tau + e^+ + \nu_e \\ &\tau^- \to \nu_\tau + \text{hadrons}\ . \end{aligned} \tag{14}$$

The first sequence, Eqs. 13 would lead to anomalous muon events

$$e^+ + e^- \to \mu^\pm + \text{hadrons} + \text{missing energy} \tag{15}$$

and the second, Eqs. 14 would lead to anomalous electron events

$$e^+ + e^- \to e^\pm + \text{hadrons} + \text{missing energy} \quad (16)$$

Anomalous muon events

The first advance beyond the e–μ events came with three different demonstrations of the existence of anomalous μ–hadron events:

$$e^+ + e^- \to \mu^\pm + \text{hadrons} + \text{missing energy.}$$

The first and very welcome outside confirmation for anomalous muon events came in 1976 from another SPEAR experiment by Cavilli–Sforza et al. (1976). This paper was entitled, "Anomalous Production of High-Energy Muons in $e^+ + e^-$ Collisions at 4.8 GeV."

I have in my files a June 3, 1976, Mark I note by Gary Feldman discussing μ events using the muon identification tower of the Mark I detector (see Fig. 7a). For data acquired above 5.8 GeV, he found the following:

> "Correcting for particle misidentification, this data sample contains 8 e–μ events and 17 μ–hadron events. Thus, if the acceptance for hadrons is about the same as the acceptance for electrons, and these two anomalous signals come from the same source, then with large errors, the branching ratio into one observed charged hadron is about twice the branching ratio into an electron. This is almost exactly what one would expect for the decay of a heavy lepton."

This conclusion was published in the paper, "Inclusive Anomalous Muon Production in e^+e^- Annihilation," by Feldman et al. (1977).

The most welcomed confirmation, because it came from an experiment at the DORIS e^+e^- storage ring, was from the PLUTO experiment. In 1977, the PLUTO collaboration published "Anomalous Muon Production in e^+e^- Annihilation as Evidence for Heavy Leptons," (Burmester et al. 1977); Fig. 8 is from that paper.

PLUTO was also a large-solid-angle detector, and so for the first time we could fully discuss the art and technology of τ research with an independent set of experimenters, with our friends Hinrich Meyer and Eric Lohrman of the PLUTO Collaboration.

With the finding of μ–hadron events, I was convinced I was right about the existence of the τ as a sequential heavy lepton. Yet there was much to disentangle: it was still difficult to demonstrate the existence of anomalous e^- hadron events, and the major hadronic decay modes

$$\tau^- \to \nu_\tau + \pi^- \quad (17)$$

$$\tau^- \to \nu_\tau + \rho^- \quad (18)$$

had to be found.

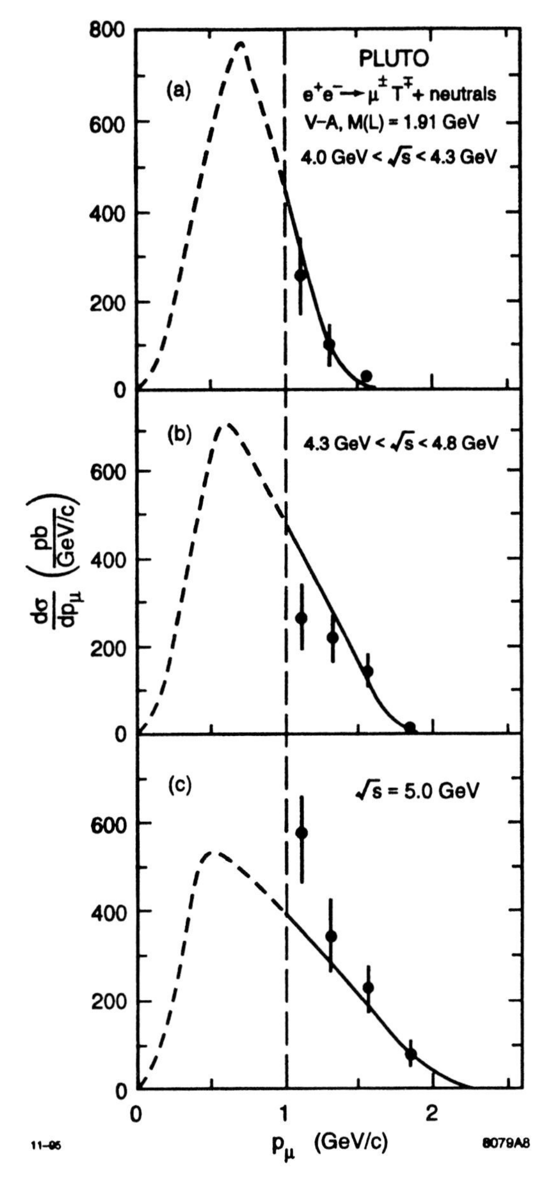

Figure 8. The momentum spectra of μ's from anomalous muon events found by the PLUTO experimenters using the DORIS e^+e^- storage ring (Burmester et al. 1977).

Anomalous electron events

The demonstration of the existence of anomalous electron events

$$e^+ + e^- \to e^\pm + \text{hadrons} + \text{missing energy}$$

required improved electron identification in the detectors. A substantial step forward was made by the new DELCO detector at SPEAR (Kirkby 1977, Bacino et al. 1978). In Kirkby's talk at the 1977 Hamburg Photon-Lepton

Conference, "Direct Electron Production Measurement by DELCO at SPEAR," he stated,

> "A comparison of the events having only two visible prongs (of which only one is an electron) with the heavy lepton hypothesis shows no disagreement. Alternative hypotheses have not yet been investigated."

The Mark I detector was also improved by Group E from SLAC and a Lawrence Berkeley Laboratory Group led by Angela Barbaro-Galtieri; some of the original Mark I experimenters had gone off to begin to build the Mark II detector. We installed a wall of lead-glass electromagnetic shower detectors in the Mark I (see Fig. 9). This led to the important paper entitled, "Electron-Muon and Electron-Hadron Production in e^+e^- Collisions," (Barbaro-Galtieri et al. 1977b). The abstract read:

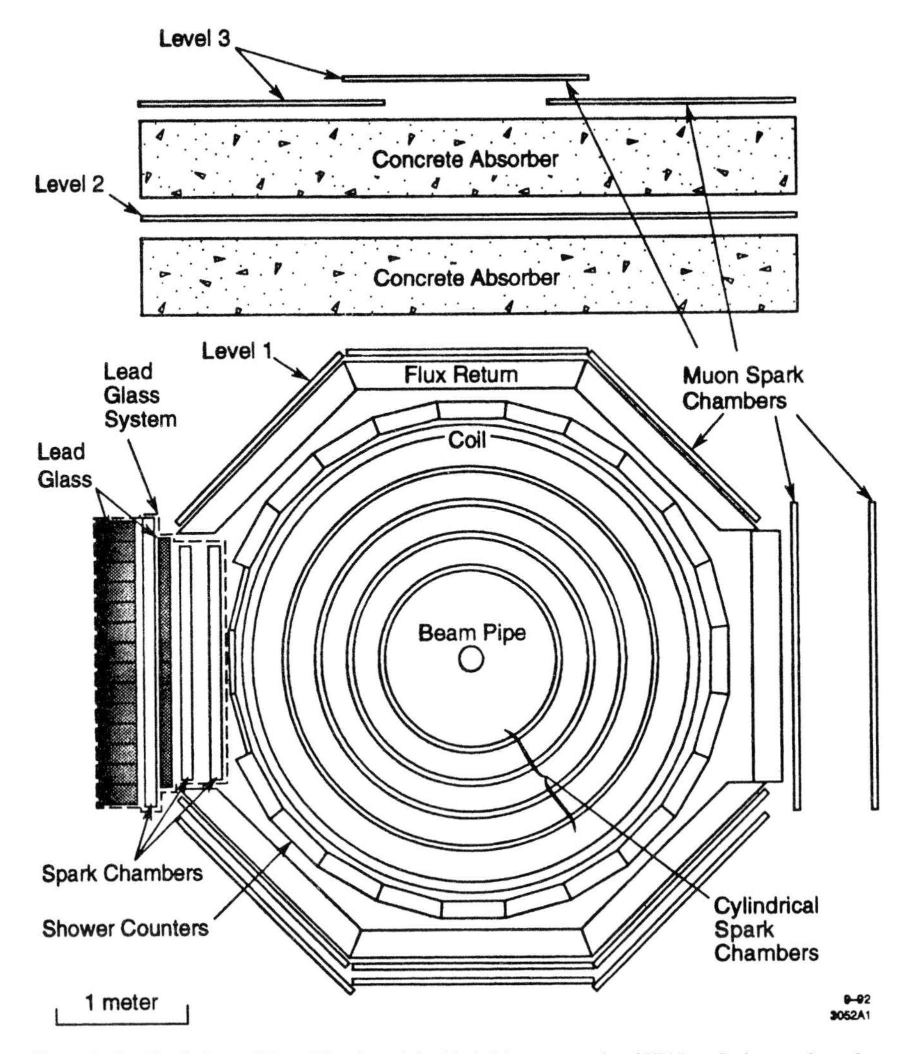

Figure 9. The "lead glass wall" modification of the Mark I detector used at SPEAR to find anomalous electron events.

"We observe anomalous e–μ and e–hadron events in $e^+ + e^- \rightarrow \tau^+ + \tau^-$ with subsequent decays of $\tau^\pm$ into leptons and hadrons. Under the assumption that they come only from this source, we measure the branching ratios $B(\tau \rightarrow e\nu_e\nu_\tau) = (22.4\pm5.5)\%$ and $B(\tau \rightarrow \text{h+neutrals}) = (45\pm19)\%$."

Semileptonic decay modes and the search for $\tau^- \rightarrow \nu_\tau \pi^-$ and $\tau^- \rightarrow \nu_\tau \rho^-$

By the time of the 1977 Photon Lepton Conference at Hamburg, I was able to report in "Review of Heavy Lepton Production in e^+e^- Annihilation," (Perl 1977) that:

"a) All data on anomalous $e\mu$, ex, ee and $\mu\mu$ events produced in e^+e^- annihilation is consistent with the existence of a mass 1.9 ± 0.1 GeV/c^2 charged lepton, the τ.
b) This data cannot be explained as coming from charmed particle decays.
c) Many of the expected decay modes of the τ have been seen. A very important problem is the existence of the $\tau^- \rightarrow \nu_\tau \pi^-$ decay mode."

The anomalous muon and anomalous electron events had shown that the total decay rate of the τ into hadrons, that is the total semileptonic decay rate, was about the right size. But if the τ was indeed a sequential heavy lepton, two substantial semileptonic decay modes had to exist: $\tau^- \rightarrow \nu_\tau \pi^-$ and $\tau^- \rightarrow \nu_\tau \rho^-$.

First, the branching fraction for

$$\tau^- \rightarrow \nu_\tau + \pi^- \tag{19a}$$

could be calculated from the decay rate for

$$\pi^- \rightarrow \mu^- + \bar{\nu}_\mu \tag{19b}$$

and was found to be

$$B(\tau^- \rightarrow \nu_\tau \pi^-) \approx 10\%. \tag{19c}$$

Second, the branching fraction for

$$\begin{aligned}\tau^- \rightarrow \nu_\tau + \rho^- &\rightarrow \nu_\tau + \pi^- + \pi^\circ \\ &\rightarrow \nu_\tau + \pi^- + \gamma + \gamma\end{aligned} \tag{20a}$$

could be calculated from the cross section for

$$e^+ + e^- \rightarrow \rho^0 \tag{20b}$$

and was found to be

$$B(\tau^- \rightarrow \nu_\tau \rho^-) \approx 20\%. \tag{20c}$$

One of the problems in the years 1977–1979 in finding the modes in Eqs. 19a and 20a was the poor efficiency for photon detection in the early detectors. If the γ's in Eq. 20a are not detected then the π and ρ modes are confused with each other. Probably the first separation of these modes was achieved using the Mark I-Lead Glass Wall detector. As reported at the Hamburg Conference by Angelina Barbaro-Galtieri (1977a)

$$B(\tau^- \to \nu_\tau \pi^-)/B(\tau^- \to \nu_\tau \rho^-) = 0.44\pm0.37.$$

Gradually, the experimenters understood the photon detection efficiency of their experiments. In addition, new detectors (such as the Mark II) with improved photon detection efficiency were put into operation. In our collaboration, the first demonstration that $B(\tau^- \to \nu_\tau \pi^-)$ was substantial came from Gail Hanson in an internal note dated March 7, 1978.

Within about a year, the $\tau^- \to \nu_\tau \pi^-$ decay mode had been detected and measured by experimenters using the PLUTO detector, the DELCO detector, the Mark I Lead-Glass Wall detector, and the new Mark II detector. These measurements were summarized (Table 4) by Gary Feldman (1978) in a review of $e^+ + e^-$ annihilation physics at the XIX International Conference on High Energy Physics. Although the average of the results in Table 4 is two standard deviations smaller than the present value of (11.1±0.2)%, the $\tau^- \to \nu_\tau \pi^-$ mode had been found.

Table 4. From Feldman (1979), the various measured branching fractions B in percent for $\tau^- \to \pi^-\nu_\tau$ in late 1978.

Experiment	Mode	Events	Background	$B(\tau \to \pi\nu)$
SLAC–LBL	$x\pi$	≈200	≈70	9.3±1.0±3.8
PLUTO	$x\pi$	32	9	9.0±2.9±2.5
DELCO	$e\pi$	18	7	8.0±3.2±1.3
Mark II	$x\pi$	142	46	8.0±1.1±1.5
	$e\pi$	27	10	8.2±2.0±1.5
Average				8.3±1.4

The year 1979 saw the first publications of $B(\tau^- \to \nu_\tau \rho^-)$. The DASP Collaboration using the DORIS $e^+ + e^-$ storage ring reported (Brandelik et al. 1979) (24±9)% and the Mark II Collaboration reported (Abrams et al. 1979) (20.5±4.1)%. Crude measurements, but in agreement with the 20% estimate in Eq. 20c. The present value is (24.8±0.2)%.

By the end of 1979, all confirmed measurements agreed with the hypothesis that the τ was a lepton produced by a known electro-magnetic interaction and, that at least in its main modes, it decayed through the conventional weak interaction. And so ends the sixteen year history, 1963 to 1979, of the discovery of the tau lepton and the verification of that discovery.

Reflections on the present and future of tau physics

Since 1979, there has been a tremendous amount of experimental and theoretical research in tau physics. There are recent reviews by Weinstein and Stroynowski (1993), Montanet (1994), and myself (Perl 1996). The proceedings of the Third Workshop on Tau Lepton Physics (Rolandi 1995) are a treasure house of information and speculation on the tau and its neutrino. There are very active experimental programs on the tau using the CESR electron-positron collider at Cornell, the LEP electron-positron collider at CERN, the SLC electron-positron collider at SLAC, and the BEPC electron-positron collider at IHEP in Beijing. This experimental research uses numbers of tau decays which are much larger than the numbers that were available during the discovery years–1000 to 10,000 more events. There are, in addition, active experiments at CERN and experiments in preparation at Fermilab that are designed to detect tau neutrinos and to look for oscillations between the tau neutrinos and other neutrinos.

There are two broad goals in tau research. One goal is to learn as much as we can about the expected behavior of the tau lepton and tau neutrino. The second goal, which is perhaps only a dream, is to find some unexpected behavior of the tau lepton, behavior that will lead us to a deeper understanding of elementary particles and basic forces. The tau is a fine candidate for such speculative research because the tau and the tau neutrino are the only particles in the third family that can be examined in a pure, isolated state. Remember that the electron-muon puzzle which set all this in motion is still not solved. The electron-muon puzzle has expanded into the electron-muon-tau puzzle. We still do not know why there are three charged leptons or understand the ratios of their masses.

In the future, there will be another increase by a factor of 10 to 100 of the number of recorded tau decays. This increase will be achieved with the high luminosity B-factories now being constructed at SLAC and KEK, and by further increases in the luminosity of the CESR electron-positron collider. And there is the special hope that a Tau-Charm Factory will be constructed at the Institute for High Energy Physics in Beijing.

I am fortunate that a short time ago some SLAC colleagues and I were able to join the CLEO Collaboration which uses the CESR collider, and so I am continuing to work on tau lepton physics. I don't have any original ideas for tau research, but I do know that the only way I get ideas is to work experimentally on a subject.

My final remark to young women and men going into experimental science is that they should pay little attention to the speculative physics ideas of my generation. After all, if my generation has any really good speculative ideas, we will be carrying these ideas out ourselves.

REFERENCES

Abrams, G. S., et al., 1979, Phys. Rev. Lett. *43*, 1555.
Alles-Borelli, V., et al., 1970, Lett. Nuovo Cimento *IV*, 1156.
Bacino, W, et al.,. 1978, Phys. Rev. Lett. *41*, 13.
Barbaro-Galtieri, A., 1977a, Proceedings of the 1977 International Symposium on Lepton and Photon Interactions at High Energies, edited by F. Gutbrod (Hamburg, 1977), p. 21.
Barbaro-Galtieri, A., et al., 1977b, Phys. Rev. Lett. *39*, 1058.
Barber, W. C, et al., 1966, Phys. Rev. Lett. *16*, 1127.
Barna, A., et al., 1968, Phys. Rev. *173*, 1391.
Bernardini, M., 1973, Nuovo Cimento *17A*, 383.
Bjorken J. D., and C. H. Llewellyn-Smith, 1973, Phys. Rev. *D7*, 887.
Brandelik, R., et al., 1979, Z. Physik *C1*, 233.
Braunstein, T., 1972, Phys. Rev. *D6*, 106.
Burmester, J., et al., 1977, Phys. Lett. *68B*, 297.
Cabibbo, N. and R. Gatto, 1961, Phys. Rev. *124*, 1577.
Camilleri, L., et al., 1969, Phys. Lett. *23*, 153.
Cavalli-Sforza, M. et al., 1976, Phys. Rev. Lett. *36*, 558.
Charpak, G., et al., 1962, Phys. Lett. *1*, 16.
Damouth, D. E., L. W. Jones, and M. L. Perl, 1963, Phys. Rev. Lett. *11*, 287.
Danby, G, et al., 1962, Phys. Rev. Lett. *9*, 36.
Ellsworth, R. W., et al., 1960, Phys. Rev. *165*, 1449.
Feldman, G. J., et al., 1977, Phys. Rev. Lett. *38*, 117.
Feldman, G. J., 1978, Proceedings of the XIX International Conference on High Energy Physics, edited by S. Hounma, M. Kawaguchi, and H. Miyazawa (Tokyo, 1978).
Feldman, G. J., et al., 1982, Phys. Rev. Lett. *48*, 66.
Hanson, G., 1978, SLAC–LBL Collaboration Internal Note, March 7, 1978.
Howard, F. T., 1967, Proceedings of the Sixth International Conference on High Energy Accelerators, edited by R. A. Mack (Cambridge, 1967), p. B43.
Kirkby, J., 1977, Proceedings of the 1977 International Symposium on Lepton and Photon Interactions at High Energies, edited by F. Gutbrod (Hamburg, 1977), p. 3.
Kirkby, J., and J. A. Rubio, 1992, Particle World *3*, 77.
Kostoulas, I., et al., 1974, Phys. Rev. Lett. *32*, 489.
Kreisler, M. N., et al., 1966, Phys. Rev. Lett. *16*, 1217.
Lai, K. W., L. W. Jones, and M. L. Perl, 1961, Proceedings of the 1960 International Conference on Instrumentation in High-Energy Physics, edited by C. E. Mauk, A. H. Rosenfeld, and R. K. Wakerling (Interscience, New York, 1961), p. 186.
Larsen, R. M., et al., 1971, SLAC Proposal SP–2.
Lipmanov, E. M, 1964, JETP *19*, 1291.
Low, F. E., 1965, Phys. Rev. Lett. *14*, 238.
Montanet, L., et al., 1994, "Review of Particle Properties," in Phys. Rev. *D50*, 1173.
Okun, L. B., 1965, Soviet Phys. JETP *20*, 1197.
O'Neil, G. K., et al., 1958, HEPL Report RX–1486.
Orito S., et al., 1974, Phys. Lett. *48B*, 165.
Perl, M. L., I. I. Rabi, and B. Senitzky, 1955, Phys. Rev. *97*, 838.
Perl, M. L., L. W. Jones, and C. C. Ting, 1963, Phys. Rev. *132*, 1252.
Perl, M. L., 1971, Physics Today, July, p. 34.
Perl, M. L., 1975a, Proceedings of the Canadian Institute of Particle Physics, Summer School, edited by R. Heinzi and B. Margolis (McGill University, Montreal, 1975).
Perl, M. L., et al., 1975b, Phys. Rev. Lett. *35*, 1489.
Perl, M. L., 1977, Proceedings of the 1977 International Symposium on Lepton and Photon Interactions at High Energies, edited by F. Gutbrod (Hamburg, 1977), p. 145.
Perl, M. L., and P. Rapidis, 1974, SLAC–PUB–1496, unpublished.
Perl, M. L., and L. V. Beers, 1994, Proceedings of the Workshop on Tau-Charm Factory in the Era of B-Factories and CESR, SLAC–451, edited by L. V. Beers and M. L. Perl (SLAC, Stanford, 1994).

Perl, M. L., 1996, Reflections on Experimental Science (World Scientific, Singapore, 1996).
Reines, F,. and C. L. Cowan, Jr., 1953, Phys. Rev. *92*, 830.
Ritson, D., S. Berman, A. Boyarski, F. Bulos, E. L. Garwin, W. Kirk, B. Richter, and M. Sands, 1964, Proposal for a High-Energy Electron-Positron Colliding-Beam Storage Ring at the Stanford Linear Accelerator Center (SLAC, Stanford, 1964).
Rothe, K. W., and A. M. Wolsky, 1969, Nuc. Phys. *B10*, 241.
Thacker H. B., and J. J. Sakurai, 1971. Phys. Lett. *36B*, 103.
Toner, W. T. et al., 1972, Phys. Lett. *36B*, 251.
Tsai, Y.-S., 1971, Phys. Rev. *D4*, 2821.
Tsai, Y.-S., and A. C. Hearn, 1965, Phys. Rev. *140B*, 721.
Tsai, Y.-S., and V. Whitis, 1966, Phys. Rev. *149*, 1348.
Voss, G., 1994, Proceedings of the International Conference on the History of Original Ideas and Basic Discoveries in Particle Physics, edited by H. Newman (Sicily, 1994).
Weinstein, A. J., and R. Stroynowski, 1993, Ann. Rev. Nucl. Part. Sci. *43*, 457.
Wilkinson, D., and H. R. Crane, 1963, Phys. Rev. *130*, 852.
Zel'dovich, Ya. B., 1963, Soviet Phys. Uspekki *5*, 931.

Frederick Reines

FREDERICK REINES

I was born in Paterson, New Jersey on March 16, 1918, the youngest of four children. My parents, Israel and Gussie (Cohen), had met and married in New York City after emigrating to the United States from the same small town in Russia. A paternal relative in Russia, the Rabbi Isaac Jacob Reines (1839–1915), was famous for his role in founding the Religious Zionist movement, Mizrachi. Manually very skilled and to some extent a frustrated machinist, my father worked as a weaver before World War I, started a silk mill business after the war, and eventually moved to Hillburn, New York, where he ran a general store. My early childhood memories center around this typical American country store and life in a small American town, including 4th of July celebrations marked by fireworks and patriotic music played from a pavilion bandstand. As a child, I enjoyed building things and participating in group singing in school. Music, and singing in particular, was to become a central lifelong interest of mine. The first stirrings of interest in science that I remember occurred during a moment of boredom at religious school, when, looking out of the window at twilight through a hand curled to simulate a telescope, I noticed something peculiar about the light; it was the phenomenon of diffraction. That began for me a fascination with light.

My early education was strongly influenced by my older siblings. Our home had many books due principally to the educational interests of my sister and two brothers, all of whom where serious students engaged in professional studies; my sister became a doctor of medicine and my brothers became lawyers. Among my activities was membership in the Boy Scouts; I rose each year through the ranks, eventually achieving the rank of Eagle Scout and undertaking leadership roles in the organization. My scientific interests also blossomed during this time in the Boy Scouts, where I began to build crystal radios "from scratch." By this time the family had returned to New Jersey, and I was a student at Union Hill High School in North Bergen. In school, I was intitially more attracted to literary interests and did not do as well in science studies. However, by my junior and senior years in high school this situation turned around sufficiently to point me in the direction of science. I was strongly encouraged by a science teacher who took an interest in me and presented me with a key to the laboratory to allow me to work whenever I wanted. I also served as Editor-in-Chief of the high school year book. In response to the year book query to students about their principal ambition, my entry was: "To be a physicist extraordinaire."

When time arrived to select a college for study in science or engineering,

I initially aimed to go to MIT, and was accepted and advised to apply for a scholarship based on my grades. However, I had a chance encounter with an admissions officer of Stevens Institute of Technology, who so impressed me by his erudition and enthusiasm for the school that I changed course and entered Stevens Institute. There, in addition to engineering studies, I participated in the dramatic society and in a dance group performance. But the college activity that I engaged in which was to have a long-standing attraction to me was singing in the chorus, where I performed solo roles in major pieces, including Handel's "Messiah." My voice and ear for music were sufficiently highly regarded that I was encouraged by the leader of the chorus to take lessons with a well-known voice coach at the Meatropolitan Opera. Since, as a student, I could not afford to pay for lessons, they were eventually provided to me free of charge. Between college and graduate school, I even thought briefly about pursuing a professional singing career, but ultimately decided against it.

The interests in music and drama that I developed in college have persisted throughout my life. Years later, while working in Los Alamos, I sang solos with the town chorus and performed with the dramatic society; my dramatic roles included the lead role in "Inherit the Wind." I also sang in performances of Gilbert and Sullivan operettas in Los Alamos. My discovery of Gilbert and Sullivan had also occurred while I was in college, and I have enjoyed occasionally entertaining colleagues and friends with G & S lyrics. The peak of my musical endeavors occurred during the period I lived in Cleveland, when I performed with the chorus of the Cleveland Symphony Orchestra under the direction of Robert Shaw and orchestra conductor George Szell.

I received my undergraduate degree in engineering in 1939 and a Master of Science degree in mathematical physics in 1941 at Steven Institute of Technology. It was during this period in 1940, that I married Sylvia Samuels. We have two children, Robert G., who currently lives in Ojo Sarco, New Mexico, and Alisa K. Cowden, of Trumansburg, New York, and six grandchildren.

I continued with graduate studies at New York University, where I worked for a time in experimental cosmic ray physics under the direction of S.A. Korff, and wrote a theoretical Ph.D thesis on "The Liquid Drop Model for Nuclear Fission" under R.D. Present. Even before completing my thesis in 1944, I was recruited as a staff member under Richard Feynman in the Theoretical Division at the Los Alamos Scientific Laboratory, to work on the Manhattan Project. During my participation in the Manhattan Project and subsequent research at Los Alamos, encompassing a period of fifteen years, I worked in the company of perhaps the greatest collection of scientific talent the world has ever known. About a year after I arrived I became a Group Leader in the Theoretical Division and, later, the director of Operation Greenhouse, which consisted of a number of Atomic Energy Commission experiments on Eniwetok atoll. In addition to my work on the results of bomb tests conducted at Eniwetok, Bikini and the testing grounds in Nevada, I directed my efforts during this period to the basic understanding of the effects of nuclear blasts, including a study of the air blast wave coauthored with John von Neumann. In 1958, I was a delegate to the Atoms for Peace conference in Geneva.

In 1951, I took a sabbatical-in-residence from my duties at Los Alamos to think about the physics I would pursue in the coming years. It was during this time that I decided to attempt the observation of the neutrino. The idea of searching for the elusive neutrino had, in fact, occurred to me as early as 1947, but the opportunity did not present itself. I was now determined to do it, and formed an extremely fruitful collaboration with Clyde Cowan, another Los Alamos staff member. We initially considered the use of a nuclear bomb test as the source of neutrinos, but soon decided that the reactor at Hanford, Washington, would be better. After the first hints of a result at Hanford in 1953, we were informed by John Wheeler about the new Savannah River reactor facility being built in South Carolina. The conditions at Savannah River were ideal for this experiment and, in 1955, Cowan and I transferred the operation there. In 1956 we observed the electron antineutrino. Shortly after that, Cowan left Los Alamos and our collaboration came to a natural end. I turned my attention for a while to gamma ray astronomy and soon began the first in a continous series of experiments at the Savannah River site to study the properties of the neutrino.

I left Los Alamos in 1959 to become Professor and Head of the Department of Physics of the (then) Case Institute of Technology in Cleveland, Ohio. During my seven years at Case, I built a group working in reactor neutrino physics, double beta decay, electron lifetime studies, searches for nucleon decay, and a very ambitious experiment in a gold mine in South Africa that made the first observation of the neutrinos produced in the atmosphere by cosmic rays. The primary goals of the experimental program were elucidation of the properties of the neutrino and probing of the limits of fundamental symmetry principles and conservation laws, such as the conservation of charge, baryon number and lepton number. Most of these experiments required the reduction of the cosmic ray muon flux in order to be successful, and the group necessarily became expert in the operation of deep underground laboratories. The projects also drew us into developing innovative detector techniques, including the use of large liquid scintillator and water Cherenkov detectors.

This line of research continued when I went, and brought my research group with me, to the new University of California, Irvine campus in 1966 to become the founding Dean of the School of Physical Sciences. I served as Dean until 1974, when I stepped down to return to full time teaching and research. I was appointed Distinguished Professor of Physics at UCI in 1987 and became Professor Emeritus in 1988. I have also served as Professor of Radiological Siences in the College of Medicine at UCI. The "Neutrino Group" at Irvine has been actively involved in a wide range of neutrino and elementary particle physics experiments, including its role in the IMB (Irvine-Michigan-Brookhaven) proton decay experiment. This group has continued the program of reactor neutrino experiments, has been the first to observe double beta decay in the laboratory, and was awarded the 1989 Bruno Rossi prize in High Energy Astrophysics by the American Astronomical Society for its joint observation (with the Kamiokande Experiment in Japan) of neutrinos from supernova 1987A. The detection of the supernova neutrinos was a particularly gratifying outcome of the IMB experiment. Our group had always been aware of the possibility of seeing neutrinos from stellar collapse in our large detectors, and several of the previous detectors had been adorned with signs identifying each of them as a "Supernova Early Warning System."

Over the years, a number of other intriguing experimental ideas and areas of investigation have been the objects of my attention, and I have devoted some time and effort to exploring the inherent possibilities. These include: the search for relic neutrinos; the"neutrino Mössbauer effect", in which a photon is replaced by a neutrino; the measurement of the gravitational constant, G, the most poorly measured non-nuclear fundamental constant by several orders of magnitude; a spherical lens space telescope; attempting to set more stringent limits on violation of the Pauli Exclusion Principle; exploration of the brain using ultra-sound; and a variety of new detector ideas. These scientific concepts, goals and challenges continue to excite and stimulate my interest.

HONORS AND AWARDS

Sigma Xi, 1944
TBπ
Centennial Lecturer, University of Maryland, 1956
Fellow of the American Physical Society, 1957
Guggenheim Fellow, 1958–1959
Alfred P. Sloan Fellow, 1959–1963
Fellow, American Academy of Arts and Sciences, 1966
Honorary Doctor of Science Degree, University of Witwatersrand, Johannesburg, South Africa, 1966
Phi Beta Kappa, 1969
Stevens Honor Award, 1971
Distinguished Faculty Lecturer, University of California, Irvine, 1979
Fellow, American Association Advancement of Science, 1979
National Academy of Sciences, 1980
J. Robert Oppenheimer Memorial Prize, 1981
Honorary Doctor of Engineering, Stevens Institute of Technology, 1984
Medal for Outstanding Reasearch, University of California, Irvine, 1985
National Medal of Science, 1985
L.I. Schiff Memorial Lecturer, Stanford University, 1988
Albert Einstein Memorial Lecturer, Israel Academy of the Siences and Humanities, Jerusalem, 1988
Bruno Rossi Prize, American Astronomical Society, 1989
Michelson-Morley Award, 1990
Goudsmidt Memorial Lecturer, 1990
New York University Plaque, 1990

Distinguished Alumnus Award, New York University, Faculty of Arts and Sciences, 1990
W.K.H. Panofsky Prize, 1992
The Franklin Medal, awarded by the Benjamin Franklin Institute Committee on Science and the Arts, 1992
Foreign Member, Russian Academy of Siences, 1994

THE NEUTRINO: FROM POLTERGEIST TO PARTICLE

Nobel Lecture, December 8, 1995

by

FREDERICK REINES

Physics Department, University of California, Irvine, California 92717, USA

The Second World War had a great influence on the lives and careers of very many of us for whom those were formative years. I was involved during, and then subsequent to, the war in the testing of nuclear bombs, and several of us wondered whether this man-made star could be used to advance our knowledge of physics. For one thing this unusual object certainly had lots of fissions in it, and hence, was a very intense neutrino source. I mulled this over somewhat but took no action.

Then in 1951, following the tests at Eniwetok Atoll in the Pacific, I decided I really would like to do some fundamental physics. Accordingly, I approached my boss, Los Alamos Theoretical Division Leader, J. Carson Mark, and asked him for a leave in residence so that I could ponder. He agreed, and I moved to a stark empty office, staring at a blank pad for several months searching for a meaningful question worthy of a life's work. It was a very difficult time. The months passed and all I could dredge up out of the subconscious was the possible utility of a bomb for the direct detection of neutrinos. After all, such a device produced an extraordinarily intense pulse of neutrinos and thus the signals produced by neutrinos might be distinguishable from background. Some handwaving and rough calculations led me to conclude that the bomb was the best source. All that was needed was a detector measuring a cubic meter or so. I thought, well, I must check this with a real expert.

It happened during the summer of 1951 that Enrico Fermi was at Los Alamos, and so I went down the hall, knocked timidly on the door and said, "I'd like to talk to you a few minutes about the possibility of neutrino detection." He was very pleasant, and said, "Well, tell me what's on your mind?" I said, "First off as to the source, I think that the bomb is best." After a moment's thought he said, "Yes, the bomb is the best source." So far, so good! Then I said, "But one needs a detector which is so big. I don't know how to make such a detector." He thought about it some and said he didn't either. Coming from the Master that was very crushing. I put it on the back burner until a chance conversation with Clyde Cowan. We were on our way to Princeton to talk with Lyman Spitzer about controlled fusion when the airplane was grounded in Kansas City because of engine trouble. At loose ends we wandered around the place, and started to discuss what to do that's interesting in physics. "Let's do a real challenging problem," I said. He said, "Let's work on positronium." I said, "No, positronium is a very good thing but Martin Deutsch has that sewed-up. So let's not work on positronium." Then I said, "Clyde let's work on the neutrino." His immediate response was, "GREAT IDEA." He knew as little about the neutrino as I did but he was a good experimentalist with a sense of derring do. So we shook hands and got off to working on neutrinos.

Need for Direct Detection

Before continuing with this narrative it is perhaps appropriate to recall the evidence for the existence of the neutrino at the time Clyde and I started on our quest. The neutrino of Wolfgang Pauli[1] was postulated in order to account for an apparent loss of energy-momentum in the process of nuclear beta decay. In his famous 1930 letter to the Tübingen congress, he stated: "I admit that my expedient may seem rather improbable from the first, because if neutrons[1] existed they would have been discovered long since. Nevertheless, nothing ventured nothing gained... We should therefore be seriously discussing every path to salvation."

All the evidence up to 1951 was obtained "at the scene of the crime" so to speak since the neutrino once produced was not observed to interact further. No less an authority than Niels Bohr pointed out in 1930[2] that no evidence "either empirical or theoretical" existed that supported the conservation of energy in this case. He was, in fact, willing to entertain the possibility that energy conservation must be abandoned in the nuclear realm.

However attractive the neutrino was as an explanation for beta decay, the proof of its existence had to be derived from an observation at a location other than that at which the decay process occurred–the neutrino had to be observed in its free state to interact with matter at a remote point.

It must be recognized, however, that, independently of the observation of a free neutrino interaction with matter, the theory was so attractive in its explanation of beta decay that belief in the neutrino as a "real" entity was general. Despite this widespread belief, the free neutrino's apparent undetectability led it to be described as "elusive, a poltergeist."

So why did we want to detect the free neutrino? Because everybody said, you couldn't do it. Not very sensible, but we were attracted by the challenge. After all, we had a bomb which constituted an excellent intense neutrino source. So, maybe we had an edge on others. Well, once again being brash, but nevertheless having a certain respect for certain authorities, I commented in this vein to Fermi, who agreed. A formal way to make some of these comments is to say that, if you demonstrate the existence of the neutrino in the free state, i.e. by an observation at a remote location, you extend the range of applicability of these fundamental conservation laws to the nuclear realm. On the other hand, if you didn't see this particle in the predicted

[1] When the neutron was discovered by Chadwick, Fermi renamed Pauli's particle the "neutrino".

range then you have a very real problem.

As Bohr is reputed to have said, "A deep question is one where either a yes or no answer is interesting." So I guess this question of the existence of the "free" neutrino might be construed to be deep. Alright, what about the problem of detection? We fumbled around a great deal before we got to it. Finally, we chose to look for the reaction $\bar{\nu}_e + p \rightarrow n + e^+$. If the free neutrino exists, this inverse beta decay reaction has to be there, as Hans Bethe and Rudolf Peierls recognized, and as I'm sure did Fermi, but they had no occasion to write it down in the early days. Further, it was not known at the time whether $\bar{\nu}_e$ and ν_e were different. We chose to consider this reaction because if you believe in what we today call "crossing symmetry" and use the measured value of the neutron half life then you know what the cross section has to be–a nice clean result. (In fact, as we learned some years later from Lee and Yang, the cross section is a factor of two greater because of parity nonconservation and the handedness of the neutrino.) Well, we set about to assess the problem of neutrino detection. How big a detector is required? How many counts do we expect? What features of the interaction do we use for signals? Bethe and Peierls in 1934[3], almost immediately after the Fermi paper on beta decay[4], estimated that if you are in the few MeV range the cross section with which you have to deal would be ~ 10^{-44} cm^2. To appreciate how minuscule this interaction is we note that the mean free path is ~ 1000 light years of liquid hydrogen. Pauli put his concern succinctly during a visit to Caltech when he remarked: "I have done a terrible thing. I have postulated a particle that cannot be detected." No wonder that Bethe and Peierls concluded in 1934 "there is no practically possible way of observing the neutrino." [3] I confronted Bethe with this pronouncement some 20 years later and with his characteristic good humor he said, "Well, you shouldn't believe everything you read in the papers."

Detection Technique

According to the Pauli-Fermi theory (1930-1934), the neutrino should be able to invert the process of beta decay as shown in Equation (1):

$$\nu + A^{Z} \leftrightarrow A^{Z-1} + e^{+} \; or \leftrightarrow A^{Z+1} + e^{-} \tag{1}$$

We chose to focus on the particular reaction

$$\nu + p \rightarrow n + e^{+} \tag{2}$$

because of its simplicity and our recognition of the possibility that the scintillation of organic liquids newly discovered by Kallmann et al.[5] might be employed on the large (~1 m^3) scale appropriate to our needs. [At the time Cowan and I got into the act, a "big" detector was only a liter or so in volume. Despite the large (> 3 orders of magnitude) extrapolation in detector size we were envisioning, it seemed to us an interesting approach worth pursuing.] The initial idea was to view a large pot of liquid scintillator with many photomultiplier tubes located on its boundary. The neutrinos would then produce positrons which would ionize causing light flashes which could be sensed by the photomultipliers and converted to electrical pulses for display and analysis.

The idea that such a sensitive detector could be operated in the close proximity (within a hundred meters) of the most violent explosion produced by man was somewhat bizarre, but we had worked with bombs and felt we could design an appropriate system. In our bomb proposal a detector would be suspended in a vertical vacuum tank in the near vicinity of a nuclear explosion and allowed to fall freely for a few seconds until the shock wave had passed (Fig. 1). It would then gather data until the fireball carrying the fission fragment neutrino source ascended skyward. We anticipated a signal consisting of a few counts assuming the predicted ($\sim 10^{-43}$ cm^2/proton) cross section, but background estimates suggested that our sensitivity could not be guaranteed for cross sections $< 10^{-39}$ cm^2/proton, four orders of magnitude short! It is a tribute to the wisdom of Los Alamos Director, Norris Bradbury, that he approved the attempt on the grounds that it would nevertheless be ~ 1000 times as sensitive as the then existing limits.

I recall a conversation with Bethe in which he asked how we proposed to distinguish a neutrino event from other bomb associated signals. I described how, in addition to the use of bulk shielding which would screen out gamma

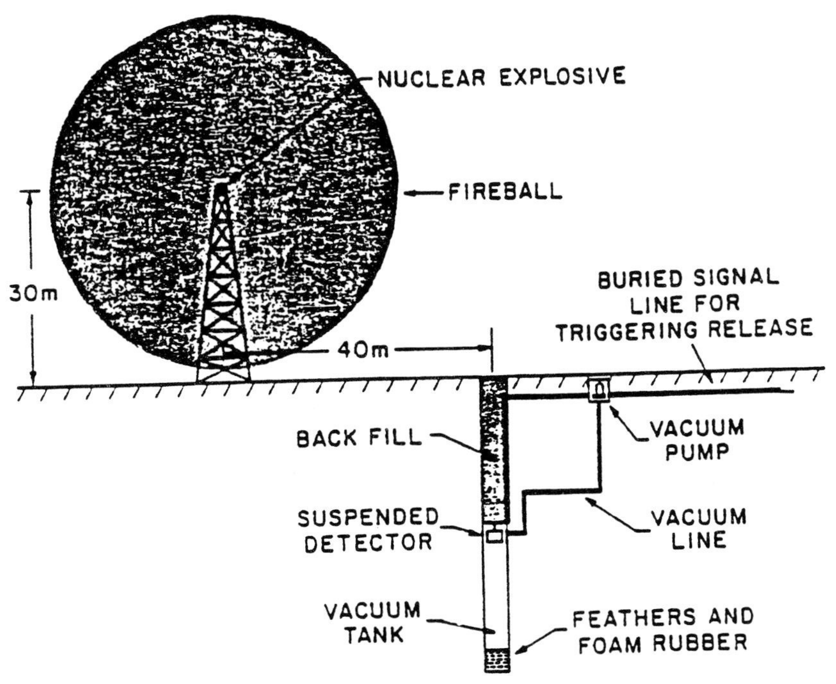

Figure 1. Sketch of the originally proposed experimental setup to detect the neutrino using a nuclear bomb. This experiment was approved by the authorities at Los Alamos but was superceded by the approach which used a fission reactor.

rays and neutrons, we could use the delayed coincidence between the product positron and neutron to identify the neutrino interaction. It was not until some months later that Clyde and I recognized this signature would drastically reduce other backgrounds so that we were able to use a steady fission reactor as a source instead of a bomb. I have wondered since why it took so long for us to come to this now obvious conclusion and how it escaped others, despite what amounted to a description of its essence as we talked to those around us. But of one thing I am certain: the open, free communication of our ideas was most stimulating to us and played a significant role in our eventual success. We were not inhibited in our communication by the concern that someone would scoop us. Neutrino detection was not a popular activity in 1952.

We sent the following letter to Fermi relating our plan to use a nuclear pile.

October 4, 1952

Dear Enrico,

We thought that you might be interested in the latest version of our experiment to detect the free neutrino, hence this letter. As you recall, we planned to use a nuclear explosion for the source because of background difficulties. Only last week it occurred to us that background problems could be reduced to the point where a Hanford pile would suffice by counting only delayed coincidences between the positron pulse and neutron capture pulse. You will remember that the reaction we plan to use is $p + \overline{\nu} \rightarrow n + \beta^{+}$. Boron loading a liquid scintillator makes it possible to adjust the mean time T between these two events and we are considering T ~ 10μsec. Our detector is a 10 cubic foot fluor filled cylinder surrounded by about 90 5819's operating as two large tubes of 45 5819's each. These two banks of ganged tubes isotropically distributed about the curved cylindrical wall are in coincidence to cut tube noise. The inner wall of the chamber will be coated with a diffuse reflector and in all we expect the system to be energy sensitive, and not particularly sensitive to the position of the event in the fluor. This energy sensitivity will be used to discriminate further against background. Cosmic ray anticoincidence will be used in addition to mercury and low background lead for shielding against natural radioactivity. We plan to immerse the entire detector in a large borax water solution for further necessary reduction of pile background below that provided by the Hanford shield.

Fortunately, the fast reactor here at Los Alamos provides the same leakage flux as Hanford so that we can check our gear before going to Hanford. Further, if we allow enough fast neutrons from the fast reactor to leak into our detector we can simulate double pulses because of the proton recoil pulse followed by the neutron capture which occurs in this case. We expect a counting rate at Hanford in our detector about six feet from the pile face of ~ $^1/_5$/min. with a background somewhat lower than this.

As you can imagine, we are quite excited about the whole business, have canceled preparations for use of a bomb, and are working like mad to carry out the ideas sketched above. Because of the enormous simplification in the experiment, we have already made rapid progress with the electronic gear and associated equipment and expect that in the next few months we shall be at Hanford reaching for the slippery particle.

We would of course appreciate any comments you might care to make.

Sincerely,

Fred Reines, Clyde Cowan

That letter elicited the response from Fermi dated Oct. 8, 1952 (Fig. 2):

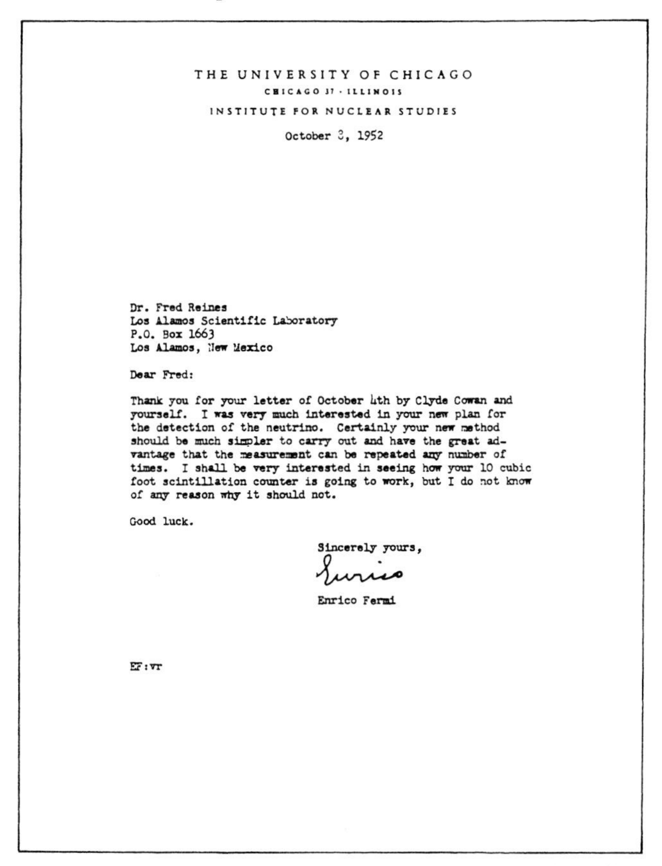

THE UNIVERSITY OF CHICAGO
CHICAGO 37 · ILLINOIS
INSTITUTE FOR NUCLEAR STUDIES

October 8, 1952

Dr. Fred Reines
Los Alamos Scientific Laboratory
P.O. Box 1663
Los Alamos, New Mexico

Dear Fred:

Thank you for your letter of October 4th by Clyde Cowan and yourself. I was very much interested in your new plan for the detection of the neutrino. Certainly your new method should be much simpler to carry out and have the great advantage that the measurement can be repeated any number of times. I shall be very interested in seeing how your 10 cubic foot scintillation counter is going to work, but I do not know of any reason why it should not.

Good luck.

Sincerely yours,

Enrico

Enrico Fermi

EF:vr

Figure 2. Letter from Fermi in response to our Oct. 4th letter to him describing our intention to use a nuclear pile rather than a bomb for the experiment.

Dear Fred:

Thank you for your letter of October 4th by Clyde Cowan and yourself. I was very much interested in your new plan for the detection of the neutrino. Certainly your new method should be much simpler to carry out and have the great advantage that the measurement can be repeated any number of times. I shall be very interested in seeing how your 10 cubic foot scintillation counter is going to work, but I do not know of any reason why it should not.

Good Luck.

Sincerely yours,

Enrico Fermi

Reflecting on the trail that took us from bomb to reactor, it is evident that it was our persistence which led us from a virtually impossible experiment to one that showed considerable promise. The stage had been set for the detection of neutrinos by the discovery of fission and organic scintillators - the most important barrier was the generally held belief that the neutrino was undetectable.

The Hanford Experiment

Our first attempt was made at one of the Hanford Engineering Works reactors in Hanford, Washington built during the Second World War to produce plutonium for the atomic bomb.

Viewed from the perspective of today's computer-controlled kiloton detectors, sodium iodide crystal palaces, giant accelerators, and several hundred-person groups, our efforts to detect the neutrino appear quite modest. In the early 1950's however, our work was thought to be large scale. The idea of using 90 photomultiplier tubes and detectors large enough to enclose a human was considered to be most unusual. We faced a host of unanswered questions. Was the scintillator sufficiently transparent to transmit its light for the necessary few meters? How reflective was the paint? Could one add a neutron capturer without poisoning the scintillator? Would the tube noise and afterpulses from such a vast number of photomultiplier tubes mask the signal? And besides, were we not monopolizing the market on photomultiplier tubes?

It soon became clear that this new detector designed for neutrinos had unusual properties with regard to other particles as well–for instance, neutron and gamma-ray detection efficiencies near 100 percent. We recognized that detectors of this type could be used to study such diverse quantities as neutron multiplicities in fission, muon capture, muon decay lifetimes, and the natural radioactivity of humans. Incidentally, the detector we designed turned out to be big enough so that a person, bent up, could fit in an insert placed in it. Intrigued, we proceeded to measure the total K^{40} radioactivity in a couple of humans. Prior to this detector development, if you wanted to measure the K^{40} in a human being you had to ash the specimen or reduce backgrounds by putting geiger counters deep underground. Incidentally, even though it was an excellent neutron as well as gamma ray detector, we resisted the temptation to be sidetracked and harvest these characteristics for anything other than the neutrino search.

Our entourage arrived at Hanford in the spring of 1953. Figure 3 shows Clyde and me sitting in front of some of our equipment. What results did we

Figure 3. Photograph of Clyde Cowan (right) and me (left) with some of the equipment we used in the Hanford experiment.

get from this particular reactor experiment? We had a 300 liter liquid scintillator viewed by 90, 2 inch photomultiplier tubes. Backgrounds were very troublesome and we found it necessary to pile and unpile hundreds of tons of lead to optimize the shielding. We worked around the clock as we struggled with dirty scintillator pipes, white reflecting paint that fell from the walls under the action of toluene based scintillator and cadmium propionate neutron capturer, etc., etc. We took the data with reactor on and off and labored until we were absolutely exhausted.

But despite our efforts, background rates due to cosmic rays and electrical noise during reactor off periods frustrated our attempts to achieve the required sensitivity.

After a few months of operation we concluded that we had done all we could in the face of an enormous reactor-independent background. We turned off the equipment and took the train back to Los Alamos.

On the way home we analyzed the data. We had checked by means of neutron sources and shielding tests that the trace of a signal, 0.4 ± 0.2 events/min., wasn't just due to reactor neutrons leaking into the detector. These marginal results merely served to whet our appetites–we figured that we had to do better than that.

Back home we puzzled over the origin of the reactor-independent signal. Was it due to "natural" neutrinos? Could it be due to fast neutrons from the nuclear capture of cosmic-ray muons? The easiest way to find out was to put the detector underground. So back at Los Alamos we performed an underground test that showed that the background was in fact from cosmic rays. While we were engaged in this background test, some theorists were rumored to be constructing a world made predominantly of neutrinos!

The Savannah River Experiment

Encouraged by the Hanford results, we considered how it might be possible to build a detector which would be even more discriminating in its rejection of background. We were guided by the fact that neutrons and positrons were highly distinctive particles and that we could make better use of their characteristics.

Figure 4 is a schematic of the detection technique used in the new experiment. An antineutrino from fission products in the reactor is incident on a water target containing cadmium chloride. As previously noted, the $\overline{\nu}_e + p$ reaction produces a positron and a neutron. The positron slows down and is annihilated with an electron, producing two 0.5 MeV gamma rays, which penetrate the water target and are detected in coincidence by two large scintillation detectors on opposite sides of the target. The neutron is slowed down by the water and captured by the cadmium, producing multiple gamma rays, which are also observed in coincidence by the two scintillation detectors. The antineutrino signature is therefore a delayed coincidence be-

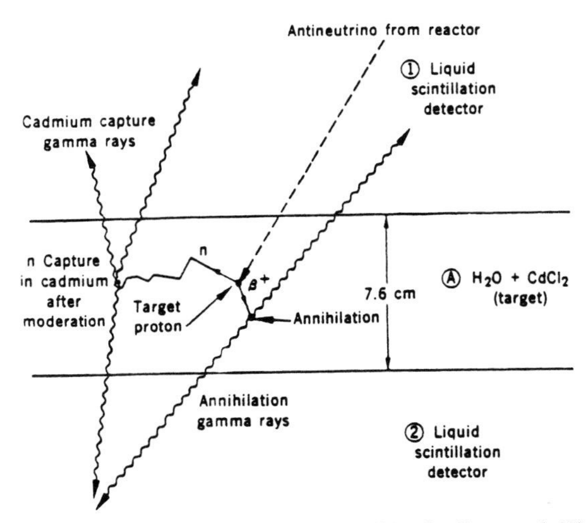

Figure 4. Schematic of the detection scheme used in the Savannah River experiment. An antineutrino from the reactor interacts with a proton in the target, creating a positron and a neutron. The positron annihilates on an electron in the target and creates two gamma rays which are detected by the liquid scintillators. The neutron slows down (in about 10 microseconds) and is captured by a cadmium nucleus in the target; the resulting gamma rays are detected in the liquid scintillators.

tween the prompt pulses produced by e^+ annihilation and those produced microseconds later by the neutron capture in cadmium.

These ideas were translated into hardware and associated electronics with the help of various support groups at Los Alamos. Figure 5 is a sketch of the equipment. It shows the target chamber in the center, sandwiched between

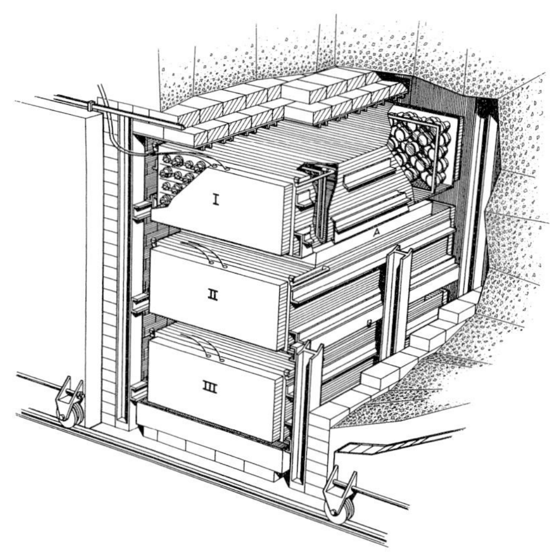

Figure 5. A sketch of the equipment used at Savannah River. The tanks marked I, II, and III contained 1400 liters of liquid scintillator solution, and were viewed on each end by 55 photomultiplier tubes. The thin tanks marked A and B were polystyrene and contained 200 liters of water, which provided the target protons and contained as much as 40 kilograms of dissolved $CdCl_2$ to capture the product neutrons.

the two scintillation chambers. Figure 6 shows one of the banks of 55 photomultiplier tubes that was used to view the scintillation chambers. Then, in the Fall of 1955 at the suggestion and with the moral support of John A. Wheeler, the detector was taken to a new, powerful (700 MW at that time), compact heavy water moderated reactor at the Savannah River Plant in Aiken, South Carolina.

Figure 6. A photograph of one of the banks of phototubes which viewed a liquid scintillator box. (See Fig. 5)

The Savannah River reactor was well suited for neutrino studies because of the availability of a well shielded location 11 meters from the reactor center and some 12 meters underground in a massive building. The high $\overline{\nu}_e$ flux, $1.2 \times 10^{13}/\text{cm}^2$ sec, and reduced cosmic ray background were essential to the success of the experiment which even under those favorable conditions involved a running time of 100 days over the period of approximately one year.

Observation of the Neutrino

At Savannah River we carried out a series of measurements to show that:[6]

a) The reactor-associated delayed coincidence signal was consistent with theoretical expectation.

b) The first pulse of the delayed-coincidence signal was due to positron annihilation.

c) The second pulse of the delayed coincidence signal was due to neutron capture.

d) The signal was a function of the number of target protons.

e) Radiation other than neutrinos was ruled out as the cause of the signal by means of an absorption experiment.

Our standard of proof was that every test must yield the anticipated result for us to conclude that we were observing the Pauli-Fermi neutrino. An unanticipated result would imply either experimental error or the need to modify our view of the neutrino.

Signal Rate

A reactor-associated correlated signal rate of 3.0 ± 0.2 events per hour was observed. This represented a very favorable set of signal to background ratios: signal to total accidental background of 4/1, signal to correlated (as in neutron capture) reactor-independent background 5/1, and signal to reactor-associated accidental background > 25/1. Determining the positron and the neutron detection efficiencies with radioactive sources and using the crudely known $\bar{\nu}_e$ flux, we found the cross-section for fission $\bar{\nu}_e$ on protons to be

$$\bar{\sigma}_{exp} = (12^{+7}_{-4}) \times 10^{-44}\ cm^2$$

compared to the expected[2]

$$\bar{\sigma}_{th} = (5 \pm 1) \times 10^{-44}\ cm^2$$

First and Second Pulses

The first pulse of the delayed coincidence pair was shown to be due to a positron by varying the thickness of a lead sheet interposed between the water target and one of the liquid scintillators, so reducing the positron detection efficiency in one of the detector triads but not in the others. The signal diminished as expected in the leaded triad but remained unchanged in the triad without lead. A further check provided by the spectrum of first pulses showed better agreement with that from a positron test source than with the background.

The second pulse was shown to be due to a neutron by varying the cadmium concentration in the target water. As expected for neutrinos, removal of the cadmium totally removed the correlated count rate, giving a rate above accidentals of 0.2 ± 0.7/hour. The spectrum of time intervals between the first and second pulses agreed with that expected for neutron capture gamma rays. A false pulse sequence in which neutrons also produced the first pulse was ruled out by use of a neutron source which showed that fast neutrons cause primarily an increase in accidental rather than correlated rates, a fact incompatible with the observed reactor-associated rates noted above.

---

[2] This was the preparity prediction.

Signal as a Function of Target Protons

The number of target protons was changed without drastically altering the detection efficiency of the system for both background and for $\overline{\nu}_e$ events. This was accomplished by mixing light and heavy water in approximately equal parts. The measured rate for the diluted target was 0.4 ± 0.1 of that for 100% $H_2 0$, a number to be compared with the expected value of 0.5.

Absorption Test

The only known particles, other than $\overline{\nu}_e$ produced by the fission process, were discriminated against by means of a gamma-ray and neutron shield. When a bulk shield measured to attenuate gamma rays and neutrons by at least an order of magnitude was added, the signal was observed to remain constant; that is the reactor-associated signal was 1.74 ± 0.12/hour with, and 1.69 ± 0.17/hour without the shield.

Telegram to Pauli

The tests were completed and we were convinced[7]. It was a glorious feeling to have participated so intimately in learning a new thing, and in June of 1956 we thought it was time to tell the man who had started it all when, as a young fellow, he wrote his famous letter in which he postulated the neutrino, saying something to the effect that he couldn't come to a meeting and tell them about it in person because he had to go out to a dance!

The message, Fig. 7, was forwarded to him at CERN, where he interrupted the meeting he was attending to read the telegram to the conferees and then

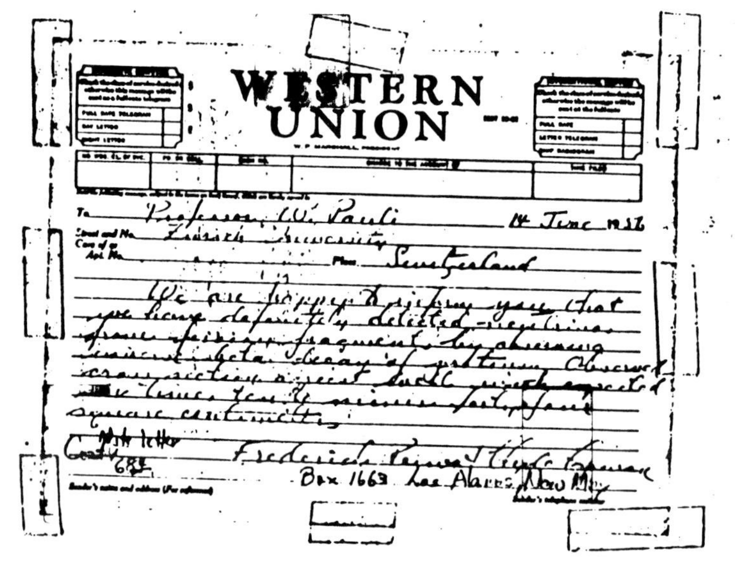

WESTERN UNION

To Professor W. Pauli 14 June 1956
Street and No. Zurich University
Place Switzerland

We are happy to inform you that we have definitely detected neutrinos from fission fragments by observing inverse beta decay of protons. Observed cross section agrees well with expected six times ten to minus forty-four square centimeters.

Frederick Reines Clyde Cowan
Box 1663 Los Alamos, New Mex

Figure 7. The telegram to Pauli which told of our detection of the neutrino at Savannah River. The contents of this message is quoted in the text.

made some impromptu remarks regarding the discovery. That message reads, "We are happy to inform you that we have definitely detected neutrinos from fission fragments by observing inverse beta decay of protons. Observed cross section agrees well with expected six times ten to minus forty four square centimeters." We learned later that Pauli and some friends consumed a case of champagne in celebration!

Many years later (~ 1986) C.P. Enz, a student of Pauli's, sent us a copy of a night letter Pauli wrote us in 1956, but which never arrived. It is shown in Fig. 8 and says, "Thanks for the message. Everything comes to him who knows how to wait. Pauli"

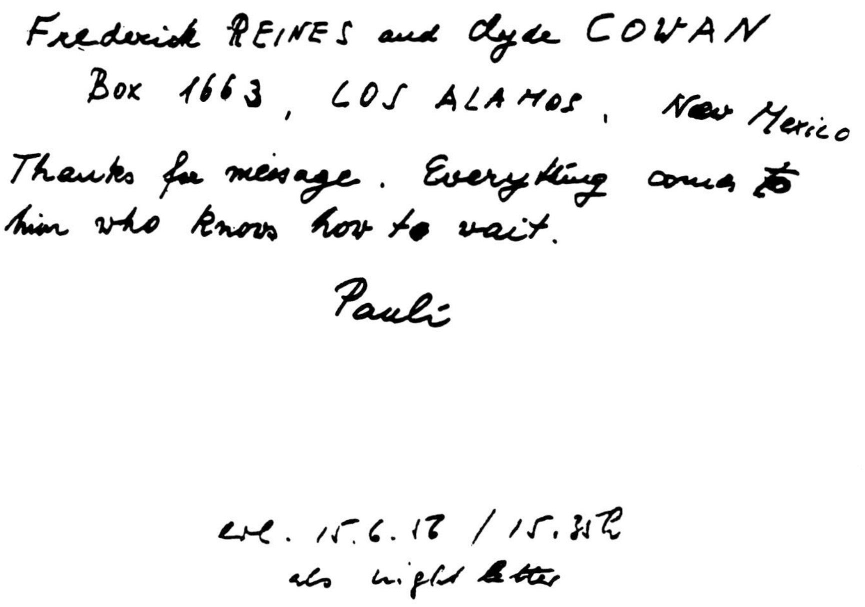

Frederick REINES and Clyde COWAN
Box 1663, LOS ALAMOS, New Mexico

Thanks for message. Everything comes to him who knows how to wait.

Pauli

als night letter

Figure 8. The night letter Pauli sent in response to our message shown in Fig. 7.

The quest was completed, the challenge met. There was, however, something missing–independent verification by other workers. As it turned out we were, in fact, correct but it took some eight years for this check to occur as a by-product of neutrino experiments at accelerators[8]. I suspect that the unseemly delay was largely due to the fact that our result was not unexpected.

Some twenty years later stimulated by the possibility of neutrino oscillations other groups also observed $\bar{\nu}_e + p$ at reactors[9].

What Next?

Having detected the neutrino the question arose, what next? What, as Luis Alvarez wrote me at the time, did we propose to do as an encore? A survey of the old notebooks indicated a variety of possibilities ranging from a study of the neutrino itself to its use as a tool in probing the weak interaction.

Neutrino-electron elastic scattering

One question I found particularly fascinating was: Did the neutrino possess a direct elastic scattering interaction with electrons

$$\bar{\nu}_e + e^- \overset{?}{\rightarrow} \bar{\nu}_e + e^- \quad (3)$$

e.g. via a magnetic moment? This question had great appeal for a variety of reasons which were not entirely sensible. First, there was no theoretical guide to suggest that such a reaction between two of nature's "simplest" particles occurred and second, reminiscent of the earlier conversation with Fermi regarding bomb neutrinos, I had no idea how to construct a suitable detector. Despite these excellent reasons for choosing a more sensible problem I decided to work on it.

The essence of the detection problem was to distinguish an electron produced by the imagined elastic scattering process from an electron produced by gamma rays or beta emitters. This sorting out of such a non-descript process occupied me, and a succession of colleagues, for some twenty years[10]. The key to the solution was the recognition (1959) that if one chose a low Z medium most of the gamma ray background would arise from Compton recoil electrons, whereas a $\bar{\nu}_e$ scattering would occur only once. It was therefore possible, in principle, to construct a detector in which spatial anticoincidences of the sequential Compton electrons would be discriminated against, thus reducing this source of background. While this idea was being translated to experimental reality and then eventual detection, various theoretical developments took place in weak interaction physics. As the theorists labored they made predictions ranging from vague qualitative guesses about magnetic moments (1934) to statements that the interaction was zero[11] (1957), that it was given by V-A (1958)[12] and that it is undefined. The situation had finally settled down by 1976 to a specific prediction with the advent of the Weinberg, Salam, Glashow theory.

That same year marked the end of our intense 20 year effort[13]. The neutrino-electron elastic scattering process has the smallest cross section of any process ever measured. The measurement also provided one of the earliest determinations of the weak mixing, or Weinberg, angle; it was only 1.2 standard deviations from the current world average.

Once again, as in the case of the inverse beta decay process, even prior to experimental verification of the elastic scattering reaction, theorists, in particular astrophysicists, assumed its existence and used it in building stellar models.

I find it interesting to contemplate the possible consequences of a closer coupling between theory and experiment in this case. If I had required a theory in the first place I would not have started to consider the scattering experiment when I did. If I had followed the theorists peregrinations I would have sacrificed the steadfastness of purpose which eventually led to the solution. This is not to say that experimentalists should proceed independently of theory, but it does suggest that the coupling should not be too tight.

Neutrino interactions with deuterons

In 1956 we also began another lengthy search; this one was for the interactions of reactor neutrinos with deuterons. In 1969 we finally observed[14] the so-called "charged-current" reaction ($\bar{\nu}_e + d \rightarrow n + n + e^+$) and in 1979[15] the "neutral-current" reaction ($\bar{\nu}_e + d \rightarrow n + n + \bar{\nu}_e$). The neutral-current reaction had been previously discovered at an accelerator at CERN in 1973 with muon neutrinos, but it was nevertheless most gratifying to see that $\bar{\nu}_e$'s exhibited the expected behaviour.

Detection of atmospheric neutrinos

In the early 1960's many authors[16] had calculated the flux of high-energy neutrinos expected to arise from the decay of K and π mesons and muons produced in the earth's atmosphere by the interaction of primary cosmic rays. A major experimental question was, How does one detect these atmospheric neutrinos? The only practical method seemed to be to detect the muons produced by the neutrinos in one of their rare interactions with matter. But this meant that one would have to place a detector deep underground to reduce the major background, the flux of muons produced directly in the atmosphere.

So in 1963 we started construction of a detector some 2 miles underground in the East Rand Proprietary gold mine near Johannesburg, South Africa. The design and construction of what was then the world's largest particle detector–a 180 ft. long, 20 ton segmented scintillation detector array–took a surprisingly short time, about one year. This experiment was a collaboration between Case Institute of Technology, Cleveland, Ohio (now Case-Western Reserve University) and the University of the Witwatersrand, Johannesburg.

On February 23, 1965, the first "natural" (meaning it did not arise from a man-made nuclear reactor) neutrino was discovered. In all, 167 such events were recorded.

Neutrino stability and oscillations

When we first turned on our detector at Savannah River in the Fall of 1955 no signals were observed. As we checked our apparatus, a desperate thought crossed our minds: the neutrino might be emitted from fission but did not survive the 11 meter journey from the reactor to our detector. Perhaps the neutrino was unstable! A moment of excitement ensued until we made some adjustments in our apparatus and neutrino-like signals began to appear. The consequence of these errors resulted in a notebook entry which suggested making a check of the inverse square law dependence of the neutrino signals on the distance from reactor to detector. But in any event we had no theoretical basis at that time for questioning the stability of the neutrino and were reminded once again that experiment was the final arbiter in these matters.

I found the idea of neutrino instability to be a "repulsive" thought but nevertheless proceeded to imagine what sorts of decay products there might be if the neutrino was, in fact, unstable. In 1974 we measured a $\bar{\nu}_e$ lifetime limit[17]. That experiment looked for the radiative decay of the neutrino at a nuclear reactor.

Early on it had been suggested by Pontecorvo and by Nakagawa et al.[18] that the neutrino may oscillate from one flavor to another as it travels from its place of origin. A graphic analogy is the change of character from dog to cat: Imagine at time zero a dog leaving his house to walk down the street to another dog house at the end of the block. As he progresses down the street a transformation takes place–his appearance gradually changes (à la Escher) from that of a dog to that of a cat! Halfway down the block the transformation is complete and the erstwhile dog–now a cat–continues on its feline journey. But the transformation goes on and, mirabile dictu, upon arrival at the dog house the erstwhile dog turned cat is once again a dog. If such bizarre behaviour is observed to occur in neutrinos it would provide evidence of the neutrino's structure. Neutrinos of all types would be construed to be built out of common building blocks whose rearrangements en route would give rise to observably different combinations.

There have been many searches for neutrino oscillations. The first experiment to report on neutrino oscillations was performed in 1979, but it was in no sense definitive; it was the same experiment in which we reported the first measurement of the neutrino-deuteron neutral current cross section[15]. Since the neutral-current reaction may be initiated by neutrinos of any flavor, whereas the charged-current reaction may be initiated only by $\bar{\nu}_e$'s, taking the ratio of the charged- to neutral-current cross sections is a sensitive test for neutrino oscillations where the oscillations occur with a wavelength short enough that the oscillation process has reached equilibrium before reaching the detector location. The results of that 1979 experiment suggested that such oscillations might occur.

Other Neutrino Physics Experiments

It must be emphasized that this grand endeavor, which we now call Neutrino Physics, is being carried out by many groups. Even in 1970 there were several such groups around the world, some using nuclear reactors, some high-energy accelerators, and others cosmic rays.

We list here only a few of the salient results that they have obtained:

In 1961 the muon-neutrino was identified in an experiment at the Brookhaven AGS[19], and this marked the beginning of the fruitful use of high-energy neutrino beams at accelerators.

In 1973, at CERN, $\bar{\nu}_\mu$-e elastic scattering was observed[20], and with it the landmark discovery of weak neutral currents.

Since the late 70's great progress has been made in studying nucleon structure functions by looking at the deep inelastic scattering of neutrinos and antineutrinos on nucleons. These studies are complementary to the deep inelastic electron and muon studies because the neutrinos couple to the nuclear constituents in a different manner and, due to parity nonconservation, they can distinguish quarks from antiquarks.

Searches for vacuum oscillations have been performed at reactors and accelerators, and since the mid 80's matter oscillations have been looked for in solar neutrinos and atmospheric neutrinos. To date there is no definitive evidence for neutrino oscillations.

Supernova 1987A was a windfall for neutrino physics[21]. Conventional supernova theory predicts that a supernova such as 1987A yields 3×10^{53} ergs (99% of its gravitational binding energy) in a burst of $\sim 10^{58}$ neutrinos in a few seconds. On earth 19 low-energy neutrino events were observed in two large Čerenkov detectors each containing several kilotons of water. All of the events were recorded within about 10 seconds; the background event rate was only a few *per day*[22]!

Many determinations of neutrino properties were extracted from the supernova data. These include neutrino mass, charge, lifetime, magnetic moment, number of flavors, etc. In addition some of the most basic elements of supernova dynamics were studied and found to be in surprisingly good agreement with predictions. One interesting consequence was the testing of the Einstein Equivalence Principle. The fact that the fermions (neutrinos) and bosons (photons) reached the Earth within 3 hours of each other provides a unique test of the equivalence principle of general relativity. The observation proved that the neutrinos and the first recorded photons are affected by the same gravitationally-induced time delay within 0.5%[23].

And while describing neutrinos arriving at the earth from the cosmos, we want to recall the intriguing history of the study of solar neutrinos. After 20 years of observation by Ray Davis and others, and now with four detectors reporting, it appears that the number of neutrinos arriving at the earth from the sun is significantly less than that expected from the standard solar model [24]. We are not yet sure whether this is telling us something about the sun or something about the properties of the neutrino.

During the latter part of the 1980's several determinations of the number of light neutrino flavors were made. The values were derived from many sources including: cosmological limits, supernova 1987A neutrinos, $p\bar{p}$ colliders, and e^+e^- colliders. By the end of the decade it was clear that there are only three families of light neutrinos[25].

Surely the longest series of experiments in neutrino physics concerns the attempt to measure the mass of the neutrino. These studies started in 1930 with Pauli's initial estimate that: "The mass of the neutron (neutrino) should be of the same order of magnitude as that of the electron and in any event no greater than 0.01 of the proton mass." Since then many techniques have been used: nuclear beta decay (especially tritium), Supernova 1987A, cosmological constraints, radiative nucleon capture of electrons and, for the mu- and tau-neutrinos, particle decays.

The Future of Neutrino Physics

The formative years of neutrino physics have been extraordinarily fruitful. But with all of the important accomplishments, are there any things left for the future? Most definitely yes.

We will continue to see more precise measurements of all of the neutrino's intrinsic properties, of course. In addition, from searches for neutrinoless double-beta decay[26] we may soon have an answer to a most fundamental question: is the neutrino Majorana or Dirac?

Also we are all anxiously awaiting the discovery of the tau neutrino, as signaled by its detection at a point remote from its origin.

There are also several outstanding issues having to do with astrophysics and cosmology. For instance: Are neutrinos an important component of the Dark Matter? And wouldn't it be exciting if someone could figure out how to observe the relic neutrinos left over from the Big Bang!

As large neutrino telescopes are constructed over the next few years, we may finally see neutrinos coming from cosmic sources such as other stars and active galactic nuclei.

I don't think it is too much to hope that we will see a resolution to the solar neutrino puzzle in the next few years. And, if we are lucky, those same detectors which will be looking for solar neutrinos may see a supernova or two.

I am confident that the future of neutrino physics will be as exciting and fruitful as the past has been.

Acknowledgements

My activities over the past 40 years could not possibly have been accomplished by myself alone. They have required the dedicated and tireless support of many talented coworkers. One in particular I must mention–my very good friend and collegue Clyde Cowan, who was an equal partner in the experiments to discover the neutrino. I regret that he did not live long enough to share in this honor with me. I also wish to thank the personnel at the Savannah River Site, and their managers, first the DuPont Corp. and later Westinghouse Corp., for their continued hospitality, interest, and good-natured tolerance over 35 years. The hospitality of Morton-Thiokol in their Fairport Mine for about 30 years is also gratefully acknowledged. The support of the U.S. Department of Energy, and its predecessors beginning with the U.S. Atomic Energy Agency, are most gratefully acknowledged.

References

[1] Pauli, W., Jr. Address to Group on Radioactivity (Tübingen, December 4, 1930) (unpublished); Rappts. Septiems Conseil Phys. Solvay, Bruxelles, 1933 (Gautier-Villars, Paris 1934).
[2] N. Bohr, J. Chem. Soc. 349 (1932).
[3] Bethe, H.A. and Peierls, R.E., Nature *133*, 532 (1934).

[4] Fermi, E., Z. Physik, *88*, 161 (1934).
[5] Kallmann, H. Phys. Rev. *78*, 62 (1950). Agena, M., Chiozotto, M., Querzoli, R., Atti Acad. Naz. Lincei Cl. Sci. Fis. Mat. Nat. Rend. *6*, 626 (1949); Reynolds, G.T., Harrison, F.B., Salvini, G., Phys. Rev. *79*, 720 (1950).
[6] Reines, F. et al., Phys. Rev. *117*, 159–173 (1960).
[7] Cowan, C.L. Jr., Reines, F., Harrison, F.B., Kruse, H.W., McGuire, A.D., Science *124*, 103 (1956).
[8] Block, M.M., Burmeister, H., Cundy, D.C., Eben, B., Franzinetti, E., Keren, J., Mollerud, R., Myatt, G., Nikolic, M., Lecourtois, A.O., Paty, M., Perkins, D.H., Ramm, C., Schultze, K., Sletten, H., Soop, K., Stump, R., Venus, W., Yoshiki, H., Phys. Lett. *12*, 281 (1964); Bienlein, J.K., Bohm, A., Von Dardel, G., Faissner, H., Ferrero, F., Gaillard, J.M., Gerber, H.J., Holm, B., Kaftanov, V., Krienen, F., Reinharz, R.A., Seiler, P.G., Staude, A., Stein, J., Steiner, H.J., Phys. Lett. *13*, 80 (1964).
[9] Reines, F., Sobel, H.W. and Pasierb, E., Phys. Rev. Lett. *45*, 1307 (1980); Boehm, F., Cavaignac, J.F., Feilitzsch, F.V., Hahn, A., Henrikson, H.E., Koang, D.H., Kwon, H., Mossbauer, R.L., Vignon, B., Vuilleumeir, J.L., Physics Letters, *97B*, 310 (1980).
[10] Reines, F., Ann. Rev. Nucl. Sci. *10*, 1 (1960).
[11] Salam, A. Nuovo Cimento, *5*, 299 (1957).
[12] Feynman, R.P. and Gell-Mann, M. Phys. Rev. *109*, 193 (1958); Marshak, R.E. and Sudarshan, G. ibid *109*, 1860 (1958).
[13] Reines, F., Gurr, H.S., Sobel, H.W., Phys. Rev. Lett. *37*, 315 (1976).
[14] Jenkins, T.L., Kinard, F.E., and Reines, F. Phys. Rev. *185*, 1599 (1969).
[15] Pasierb, E., Gurr, H.S., Lathrop, J., Reines, F., and Sobel, H.W., Phys. Rev. Lett. *43*, 96 (1979)
[16] K. Greisen, *Proceedings of the International Conference on Instrumentation for High Energy Physics, Berkeley, California, September 1960* (Interscience Publishers, Inc., New York, 1961), p. 209; M. A. Markov and I. M. Zheleznykh, Nucl. Phys. *27*, 385 (1961); G. T. Zatsepin and V. A. Kuzmin, Zh. Eksperim. i Theor. Fiz. *41*, 1919 (1961) [translation: Soviet Physics.-JETP *14*, 1294 (1962)]; R. Cowsic, Proceedings of the Eighth Internat.. Conf. on Cosmic Rays, Jaipur, India, December 1963, edited by R. R. Daniels, et al.
[17] Reines, F., Sobel, H.W., and Gurr, H.S., Phys. Rev. Lett. *32*, 180 (1974).
[18] Pontecorvo, B., Zh. Eksp. Teor. Fiz *53*, 1717 (1967) (Sov. Phys. J.E.T.P. *26*, 984 (1968)); Nakagawa, M., Okonogi, H., Sakata, S., Toyoda, A. Prog. Theor. Phys. *30*, 727 (1963).
[19] Danby, G., Gaillard, J.M., Goulianios, K., Lederman, L.M., Mistry, N., Schwartz, M., Steinberger, J., Phys. Rev. Lett. *9*, 36 (1961);
[20] Hasert, F.J. et al., Phys. Lett. *46B*, 121 (1973).
[21] Kielczewska, D., Int. Jour. of Mod. Phys. D, *3*, 331 (1994).
[22] Bionta, R.M. et al., Phys. Rev. Lett. *58*, 1494 (1987); Hirata, K. et al., Phys. Rev. Lett. *58*, 1490 (1987).
[23] Longo, M.J., Phys. Rev. Lett. *60*, 173 (1988); Krauss, L.M. and Tremaine, S., Phys. Rev. Lett. *60*, 176 (1988).
[24] Bahcall, J.N., Lande, K., Lanou, R.E., Learned, J.G., Robertson, R.G.H., and Wolfenstein, L., Nature *375*, 29 (4 May 1995).
[25] See for example: Denegri, D. and Martinelli, G., in *Neutrino physics*, K. Winter (ed.), Cambridge Monographs on Particle Physics, Nuclear Physics, and Cosmology, Cambridge University Press (1991).
[26] Moe, M. and Vogel, P., Annu. Rev. Nucl. Part. Sci. *44*, 247 (1994).